Accession no.
36175152

Fluid Pasts

WITHDRAWN

D1609998

DUCKWORTH DEBATES IN ARCHAEOLOGY

Series editor: Richard Hodges

Against Cultural Property John Carman
Archaeology: The Conceptual Challenge Timothy Insoll
Archaeology and International Development in Africa Colin Breen & Daniel Rhodes
Archaeology and Text John Moreland
Archaeology and the Pan-European Romanesque Tadhg O'Keeffe
Beyond Celts, Germans and Scythians Peter S. Wells
Bronze Age Textiles: Men, Women and Wealth Klavs Randsborg
Changing Natures: Hunter-Gatherers, First Farmers and the Modern World Bill Finlayson & Graeme M. Warren
Combat Archaeology John Schofield
Debating the Archaeological Heritage Robin Skeates
Early Islamic Syria Alan Walmsley
Ethics and Burial Archaeology Colin Breen and Daniel Rhodes
Fluid Pasts: Archaeology of Flow Matt Edgeworth
Gerasa and the Decapolis David Kennedy
Houses and Society in the Later Roman Empire Kim Bowes
Image and Response in Early Europe Peter S. Wells
Indo-Roman Trade Roberta Tomber
Loot, Legitimacy and Ownership Colin Renfrew
Lost Civilization: The Contested Islamic Past in Spain and Portugal James L. Boone
Museums and the Construction of Disciplines Christopher Whitehead
The Origins of the English Catherine Hills
Pagan and Christian: Religious Change in Early Medieval Europe David Petts
Rethinking Wetland Archaeology Robert Van de Noort & Aidan O'Sullivan
The Roman Countryside Stephen Dyson
Shipwreck Archaeology of the Holy Land Sean Kingsley
Social Evolution Mark Pluciennik
State Formation in Early China Li Liu & Xingcan Chen
Towns and Trade in the Age of Charlemagne Richard Hodges
Villa to Village Riccardo Francovich & Richard Hodges

Fluid Pasts

Archaeology of Flow

Matt Edgeworth

LIS LIBRARY
Date 09/04/13 Fund i-che
Order No 2382404
University of Chester

Bristol Classical Press

First published in 2011 by
Bristol Classical Press
an imprint of
Bloomsbury Academic
Bloomsbury Publishing Plc
50 Bedford Square
London WC1B 3DP

© 2011 by Matt Edgeworth

All rights reserved. No part of this publication
may be reproduced, stored in a retrieval system, or
transmitted, in any form or by any means, electronic,
mechanical, photocopying, recording or otherwise,
without the prior permission of the publisher.

CIP records for this book are available from the
British Library and the Library of Congress

ISBN 978-0-7156-3982-5

Typeset by Ray Davies
Printed and bound in Great Britain by
CPI Antony Rowe, Chippenham and Eastbourne

www.bloomsburyacademic.com

Contents

List of illustrations

Preface and acknowledgments

Flowing water, like air, tends to be regarded as immaterial. Anything that is fluid, anything that flows, is not usually counted as material culture, no matter how culturally shaped and manipulated it might be. Once accepted as archaeological matter in its own right, however – once incorporated into the archaeologist's way of seeing – flowing water and other kinds of material flow can radically transform the perception of past landscapes, adding another dimension to archaeological interpretation.

The idea behind the book was first sketched out in an online article, 'Rivers as Artifacts', written for *Archaeolog* in April 2008. I still believe that this is a useful way of understanding rivers. But much has happened since then and every case-study examined showed rivers to be a particularly dynamic and vibrant kind of material culture. Part wild and part artificial, the flow of rivers shapes human action just as much as it is shaped by it. It holds us in its sway even as we harness its power. Flow is a force that has had to be reckoned with and adapted to by people of all times and places across the globe. It needs to be reckoned with by archaeologists too.

Early versions of chapters were presented as papers throughout 2010 – in the MSRG (Medieval Settlement Research Group) Conference at the University of Leicester, the session on water installations at the European Archaeology Conference at Den Hague, a seminar in the series on the theme of water at the University of Birmingham, and the session on Manifestos for Materials at TAG (Theoretical Archaeology Conference), University of Bristol – with feedback from participants at all these events being very useful.

The book was mostly written at the University of Durham in Spring, 2011. I was fortunate to hold a Leonard Slater fellowship at University College while also affiliated to the Institute of Advanced Studies and the Department of Archaeology, and I would like to acknowledge the generous help afforded to me at all these centres. Thanks also to Deborah Blake at Bristol Classical Press and series editor Richard Hodges. Others who have helped in various ways include Brent Fortenberry, Lyndon Cooper, Susan Ripper, Chris Gerrard, Paul Harrison, Sarah Semple, Peter Carne, Neil Christie, Paul Courtney, Louise Rayne, Tony Wilkinson and many colleagues on the Wallingford Burh to Borough Research Project, though only I bear responsibility for the views put forward in the following pages. I dedicate this book to my late father, whose way of thinking and talking about things greatly influenced all those who knew him.

1

Rivers as entanglements of nature and culture

Introduction

Rivers move. Walk next to a river and watch its currents, whorls and eddies, changing from moment to moment. The flow of water may carry all sorts of material with it, from leaves to fallen trees to discarded human artefacts. There is a general movement from upstream to downstream, but not a uniform one. Throw sticks into various parts of the flow and watch them being carried along: these give a good indication of the speeds and directions of currents in different parts of the river, at least on the surface. Engage with the river a little more – perhaps by wading through its shallows, swimming in its deeper pools, or casting in a fishing line – and you will encounter its unseen undercurrents too.

Rivers move in many other ways. A mountain stream carves its own channels as it flows. Material taken away here will be carried (rolled along the bed, suspended in the current, or held in solution) to be deposited further downstream. In this sense the stream consists of flowing silt and pebbles and other solid matter as well as water.

In their middle to lower reaches, many rivers take the form of single channels meandering across broad floodplains comprised of sediments they themselves have laid down, shape-shifting all the while. Meanders are enlarged through erosion of outer and downstream sides of bends, and deposition of sediment on inner and upstream sides of bends, so that the whole system of meanders migrates downstream in a wave-like pattern. Occasionally movement of one meander will overrun another, leading to formation of a cut-off or chute and temporary straightening of the river channel at that point, only for the process of meander formation to begin all over again. If speeded up this would appear as a flowing motion, rather like the looping swimming action of an eel. The river would be seen to be constantly on the move, its course shifting and its floodplain taking shape.

Some active meandering rivers have been stabilised by human modification, and their channels today tend to be less dynamic and to stay longer in one place. There are other kinds of river and floodplain types too, some with multiple braided or branching channels, some with channels that are liable to sudden rather than gradual changes of course. But let us stick for the time being with the snaking meanders described above. The map in Fig. 1.1 shows the many former courses of the Mississippi in its main meander belt over the last 2,000 years or so – one of a remarkable series of fifteen coloured maps by geologist Harold Fisk (1944). It gives an idea of just how much a river can move over a relatively short period of time.

The water of the Mississippi was famously described as 'too thick to drink, too thin to plough'. When it breaks its banks or breaches its levees during a flood, much of that mud is deposited across the broad floodplain. More mud is dropped further downriver. As the river approaches the sea, the flow slackens and releases its grip on the fine particles. The mud laid down makes the river divide into multiple branches to form a fan-shaped delta, with channels shifting this way and that as yet more sediment is deposited.

But there is something missing in the way we understand and investigate rivers. We tend to see them as part of systems and cycles that have little to with archaeology as such. Rivers are regarded mostly as natural rather than cultural entities, and accordingly are taken to be the subject matter of physical geography, fluvial geomorphology, hydrology, sedimentology and ecology. They are subjected to many scientific forms of investigation – not so much to the cultural theories that might

Fig. 1.1. The entangled river. Part of the Mississippi meander belt, showing present and former courses, mapped by geologist Harold Fisk in the 1940s (source: United States Army Corps of Engineers).

be applied to landscapes, townscapes and artefacts that archaeologists normally deal with.

The problem is – it is actually very difficult to find any river in a pristine, natural state. There is a very real sense in which rivers today *are* cultural artefacts. Over the centuries, in most parts of the world, rivers and their flow have been artificially shaped, diverted, bifurcated, narrowed, shortened, widened, channelled, straightened, dredged, deepened, dammed, redirected, embanked, canalised, or modified in countless other ways. Sometimes modification is intentional: sometimes it is the unintended effect of other actions. If the river ever was entirely a natural entity, it has long since been at least partially honed to fit human projects. If it ever was wholly wild and untamed, it has long since been at least partially domesticated. And if it ever was merely an environmental entity, it has long since become part of the cultural landscape. In Chapter 2 we will return to consider in more detail some of the entangled loops depicted in Fisk's map of the Mississippi, to find that human actions are woven into – literally entangled with – the movements of the river.

Most readers will be familiar with the crucial part that rivers play in the water cycle, draining the land of water that has fallen as rain and taking it to the sea, where it evaporates into the atmosphere to eventually fall as rain once more for the cycle to begin all over again. The circular flow of water – powered by the energy of the sun and the gravity of the earth – is usually presented as a natural cycle. There is an inherent incompleteness about this idea, which is part of our *a priori* conceptual framework for understanding rivers (Linton 2008: 645). For one thing, human intervention in the water cycle has radically transformed it, so that today it is just as much a social and political cycle as it is a natural one (and how far this can be said to extend back into the past is an open question). But the rhetoric of natural cycles continues unabated.

It is not just in the natural sciences that the cultural dimension of rivers is downplayed. Much the same occurs in archaeology and other social sciences too. There is a deep polarisation inherent in our thought, language and forms of analysis which encourages us to split environments into cultural and natural components. Such polarisation can take on temporal as well as spatial forms. Thus rivers like the Columbia or the Missouri in the USA have been so intensively shaped in the past two hundred years that their status as 'organic machines' (White 1995) or 'heavily modified cyborg-like environments' (Scarpino 1997: 5) is practically incontrovertible. But many compare the acculturated state of the rivers today with the natural state they were in before the impact of colonialism and industrialisation. In other words, the river may be a cultural artefact now (the reasoning goes) but until recently it was a natural entity.

Both spatial and temporal versions of the culture/nature polarity would seem to rule out the possibility of an archaeology of rivers, which might be perceived as something of a contradiction in terms. Archaeology is the study of material culture from the past. If rivers really are natural entities – or were natural until their modern cultural transformation – then surely archaeology should focus on the more traditionally defined material culture of past human societies, leaving rivers for natural scientists to study?

That is a view which will be challenged in the following pages, where a very different proposition will be put forward. The proposition is this. *Most rivers are neither natural nor cultural, but rather entanglements of both.* It will be argued moreover that this entanglement, far from being a modern phenomenon, actually goes back much further back into the past than we might think, contradicting common-sense notions of rivers as 'natural'. Human interventions in river processes – and, for that matter, river interventions in human activities – are not specific to the contemporary and recent past. In order to properly understand rivers and their entanglement with people through time, therefore, the methods of archaeology and history need to be combined with those of hydrology, sedimentology and geomorphology. I call such a multidisciplinary approach towards rivers the 'archaeology of flow'.

Meandering streams: a historical transformation

In the 1950s, Luna Leopold and Gordon Wolman carried out their classic study on fluvial processes at work in meandering streams, basing it on empirical observations of small rivers like Brandywine River, Watts Branch and Seneca Creek in Maryland and Pennsylvania (Wolman and Leopold 1957). This account of river formation processes was accepted as a standard model, subsequently applied to the understanding of meandering rivers across the world. On the basis of their work a particular kind of stream, meandering in a single S-shaped channel through a self-formed floodplain of fine-grained sediments, with high near-vertical banks, became the archetype of a natural watercourse, untouched by human activity.

It turns out, however, that the streams studied by Leopold and Wolman are not so natural after all. A recent paper in *Science* by Robert Walter and Dorothy Merritts showed that these channels had in fact been extensively modified by prolific construction of mill dams during the colonial period. The consequent changes in flow radically altered not only the shape of the streams but the whole floodplain (Walter and Merritts 2008).

Walter and Merritts were initially puzzled by how such small and shallow streams could have deposited the several metres of sediment which constituted the floodplains. Modern flood patterns simply do not produce that amount of sediment. Where had it all come from? What they found – through an extensive programme of coring, trenching, historical research and the study of old maps as well as modern LIDAR data and aerial photos – was that the sediment had accumulated behind the colonial mill dams, the existence of which had not been taken into account in the Leopold and Wolman study.

Underneath the metres of sediment, Walter and Merritts discovered evidence for quite a different kind of river – multiple branching channels threading their way through broad forested wetlands, with lots of linked small pools and islands. This would perhaps have been the kind of environment that the early colonists encountered when they first arrived on the scene. By the eighteenth century, however, scores of dams had been built across many of the streams, often only spaced a mile or two apart, so that the valley effectively came to be stepped like a descending staircase. The dams were only up to a few metres high, but sometimes they stretched from one side of the valley to the other, turning multi-channelled streams and linked pools into a single mill-pond on the upstream side of the dam. Sediment accumulated behind the dams and gradually the mill-ponds silted up – a process hastened by rapid deforestation carried out by colonial farmers upriver, which increased erosion of soils and amount of sediment carried by the stream. As steam turbines began to replace water power, the old disused dams were breached, resulting in bursts of fast-flowing water cutting through the sediment that had built up behind the next dam downstream. It was these streams that developed into the meandering forms studied by Leopold and Wolman, who assumed them to be natural forms (summarised from Walter and Merritts 2008).

The authors of the new study state that the typical modern, incised, meandering form can now be recognised as 'an artefact of the rise and fall of mid-Atlantic streams in response to human manipulation of stream valleys for water power' (Walter and Merritts 2008: 303). The transition from multiple braided or branching channels to an active meandering form was at least partly caused by human modification. It is strongly suggested by the authors that many of the rivers and floodplains of Europe underwent similar transformations in medieval and post-medieval times, given the great density of watermills known to have existed (Walter and Merritts 2008: 304). In this respect it is interesting to compare their findings with a study of streams in the south of England (Downward and Skinner 2005).

The Walter and Merritts paper, as Montgomery (2008: 292) points out, solves a long-standing conundrum. It has long been recognised that intensive agriculture and deforestation bring about greater river erosion, increasing the amount of sediment carried by the river – yet sediment deposition from the streams into the Atlantic is known to have declined. Where was the missing sediment? The answer, it is now clear, is that much of it had been retained behind the dams. The supposed 'flood-plain' of the meandering streams in fact consisted mainly of dam-retained sediment that would otherwise have been washed down to the sea.

Although Walter and Merritts are earth scientists and not archaeologists as such, I take their work on mid-Atlantic streams to be a classic study in archaeology of flow – combining the perspectives of archaeology and history with those of hydrology and geomorphology in order to engage with the river as a mixture of natural and cultural forces. Their observations of the previously unobserved mill dams (the tops of which protruded through the sediment in some cases) were essentially archaeological observations of the material traces of past human activity which both *impacted on* and were *impacted on by* the flow of the river. Implications of the study are vast, because they go right to the very foundations of our very concept of 'natural'. The study reveals just how easy it is for assumptions about natural forces at work to hide even recent human interventions in river systems. Underlying and enfolded into the multiple layers of 'natural' sediment are artificial structures that have become an embedded part of local geomorphology.

Insights that Leopold and Wolman gained into fluvial processes in their original study still stand, but the lesson to be drawn from Walter and Merritt's revision of their work is that river systems, to be understood holistically (with regard to both natural *and* cultural dimensions, or rather the intermingling of both) need to be studied with an archaeological eye as well as a hydrological one.

Rivers as artefacts?

Most rivers can be described as artefacts in the sense that they have been shaped by human action. As clearly demonstrated by Walter and Merritts, mill dams and other river structures can have a massive impact on the geomorphology of the river. But rivers are also artefacts in the sense that they have instrumental uses; that is, although comprised partly of flowing liquid, they have cultural affordances just like solid artefacts do. The difference is that their affordances tend to be associated with flowing energy rather than form. It is flow, as much as form, that is shaped and channelled for a vast range of instrumental uses. Supply of running water in the form of a river or stream might facilitate the washing of clothes, trapping of fish, transport of logs, movement of ships, irrigation or drainage of fields, driving of water-wheels, flushing out of ponds, cooling down of heated materials, or performance of numerous other tasks. Rivers and their flow may be radically re-shaped to maximise efficiency in these respects. And like other artefacts, flowing water can be used to shape diverse kinds of materials and turn these into artefacts too. Thus water-powered mills were used at various times and places not just to grind grain but also to saw wood, to pulverise rocks, to full cloth, mint coins, sharpen swords, drive weaving looms, and so on. At least until the advent of steam power, the use and control of water flow in many parts of the world was embedded in everyday industrial processes and domestic activities – interwoven into the very fabric of economic and cultural life.

However, it has to be said that flowing liquid does not fit easily into the categories that come with the term 'artefact'. We tend to think of artefacts as made out of materials that are more or less solid, like wood, stone, bone, metal – or perhaps softer material like cloth. While liquids are materials too, their non-solidity makes them less material, less artefactual, in our eyes. Everyday notions of materiality stress the solidity of material objects, their sharp edges or solid surfaces, and the affordances these have for human action. Books about artefacts and material culture (Tilley et al. 2006; Knappett 2005; Graves-Brown 2000) tend to be about solid things like pottery, keys, cars, baskets, clothes, jewellery, and so on. Not liquids like water (though see Ingold 2007 for all that usually gets left out of archaeological accounts of materials and material culture).

Rivers can of course have solid surfaces. As the temperature drops below freezing the surface of water hardens into ice. Affordances of the surface change into those normally associated with hard and solid surfaces, then change back again as temperature rises and the ice melts. There is a shift from fluid to fixed and back to fluid again (Normark 2009).

This changing of matter from one state to another is not something normally expected of artefacts. Common sense tells us that an artefact should have a given form which, though malleable in some instances, stays more or less constant. But rivers do not conform to this at all. Consider what happens when a river floods. In breaking its banks, it loses its outlines and therefore its shape. The very boundaries of the river become fluid. From being 'formed' it becomes relatively 'formless', and the wild aspect of the river reasserts itself. When the flood recedes, the artificial edges re-emerge and the river shrinks back into its culturally applied form once again. Sometimes, then, a river is more of an artefact than at other times. The very status of the river as an artefact on the one hand or natural entity on the other is fluid. Neither one thing nor the other, it partakes of both. And these different aspects, far from being stable, can fluctuate in relation to each other. Even a flood, of course, is not entirely a natural phenomenon. Factors that can help to cause floods include 'artificial' patterns of drainage, embankment, deforestation, and so on, as well as 'environmental' factors such as climate and geology – intermeshed together as all these are.

Floods and droughts, along with shifting meanders and other river movements, present challenges to mapmakers – who prefer the edges of things to stay in one place. Defining the edge of a river is difficult at the best of times. Even when standing on the bank of a river you might think you are on dry land, when actually (unless the river bank has been artificially clad with waterproof material) part of the river may be flowing through the ground under your feet. There is no clear dividing line between a river and its floodplain.

To ask the question – 'are rivers natural entities or cultural artefacts?' – is therefore somewhat superfluous. Rivers defy categorisation as one thing or the other. If we have to classify, we might call them 'wild artefacts' – for whereas the form of most artefacts is more or less fixed, rivers have a wildness about them that cannot be entirely contained. Unlike things crafted out of stone or other solid material, a river is an artefact that can escape the bounds of its culturally applied form. The question of whether a river is an artefact or not is a variation of the nature/culture polarity already mentioned. Bruno Latour once said 'the very notion of culture is an artefact created by bracketing Nature off' (Latour 1993: 104) – to which we can add the converse, that *the very notion of nature is an artefact created by bracketing Culture off*. Rivers are viewed as natural entities only through a playing down of their cultural attributes.

There is of course a point at which water becomes almost completely subject to cultural control and material transformation – where flow itself is no longer a natural energy but an artificially generated force that can be switched on and off at will, like modern plumbing. Some archaeological examples arguably fall into that category. The wooden or lead water pipes of Cistercian abbeys, the aqueducts and fountains of ancient Rome, the conduits and cisterns of Carthage, are all examples of material traces of water systems so controlled and managed that they might be said to be cultural artefacts through and through and to have lost their wild aspects.

That is not what this book is about. The term 'water management', with its emphasis on cultural control and manipulation of the water resource, carries with it a strong implication of the passivity of water relative to human agency. In this book, however, the focus is on rivers as entanglements of natural and cultural forces – almost a kind of wrestle – where water flow is itself an active participant in the transaction. There is some attempt at control, but this is always met with river responses which are never entirely predictable, which require counter-responses, and so on. Flow is corralled, but its wildness is never entirely contained or enclosed.

This is not to downplay the cultural aspect of rivers. The following example illustrates ways in which cultural agencies may have been at work in the past even in the most 'natural' of places. It raises further questions about how we separate natural heritage from cultural heritage, when the two are so closely bound together in the first place (see Holtorf 2008).

Döda Fallet: the ruin of a river

Döda Fallet means 'dead waterfall'. Now a tourist attraction and nature reserve, it consists of the remains of a white-water rapid in a spectacular steep-sided gorge – an abandoned course of the River Indal in Sweden. The descending linear field of boulders, rock ledges, scour holes and stagnant pools, overgrown in places with thick vegetation, is redolent of former flow. It might be taken to be wholly the result of natural processes, nothing to do with archaeology at all. The place after all is classed as a nature reserve, and even appears in the Reader's Digest *Book of Natural Wonders* (1978: 127). But to suppose it to be wholly natural would be a mistake. For the Döda Fallet cannot be explained in terms of natural processes alone.

The gorge was scoured out by torrents of water from melting glaciers, at the end of the last Ice Age. A parallel channel was blocked by glacial debris, forming an esker or glacial dam which held the 16 mile long Lake Ragunda behind it. The Döda Fallet channel was at that time the main outlet of water from the lake. The rapid formed where bed material was highly resistant to water erosion, and the steep gradient caused the stream to be faster and shallower. The many rocks obstructing water flow caused turbulence and turned the churning water white as air bubbles mixed in – hence the name Storforsen, or great whitewater rapid.

So it was for thousands of years, until human activity in the form of deforestation and logging began to impact on the river. An important consideration was that logs could never survive intact the rough passage through the Storforsen rapid – they

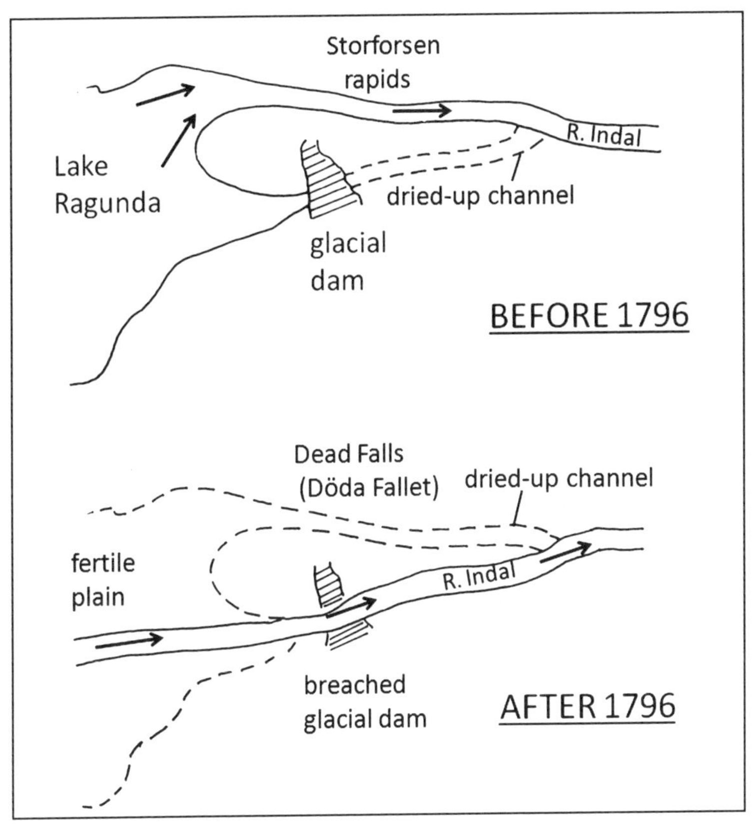

Fig. 1.2. Lake Ragunda and the Storforsen rapids, before and after 1796 (source: map in *Svenska Familj-Journalen* 1864).

had to be portaged overland past that turbulent stretch of river instead.

In the late eighteenth century a logger called Magnus Huss came on the scene. He was otherwise known as 'Wild Huss' (individuals engaged in trying to tame rivers often acquire, it seems, something of the wildness of rivers in return). In 1794 he started work on excavating a canal through the natural dam of glacial material, in an attempt to bypass the Storforsen rapid and make use of the parallel channel for moving logs.

Fig. 1.3. The Döda Fallet, Sweden (photo: Joel Torsson, 2006, CC by 3.0).

Unfortunately the unconsolidated sands and gravels that formed the dam, destabilised by the digging of the canal, suddenly gave way during unexpectedly high Spring floods in 1796. The waters of Lake Ragunda poured through the gap, scouring and deepening the canal into part of a new river course, linking up with the old parallel channel and abandoning completely the old channel. In a single night the lake was drained, causing immense damage to farms, houses, forests and land as the torrent of water swept downstream. The great whitewater rapid was transformed into the 'ruin' of a waterfall in the abandoned channel. Now all that flows here are the tourists along the many wooden walkways.

This story of human-river entanglement did not end there, however. Long-term effects of that human intervention and the river's response in 1796 were huge, with repercussions even today. The landscape was transformed upstream and downstream of the point of engagement. Downstream, some of the enormous amount of sediment eroded and swept downriver by the floodwater was re-deposited on the Indal delta in the Baltic Sea, forming new land on which Sundsvall-Midlanda Airport was later constructed. Upstream, the drained Lake Ragunda became a fertile plain, on which the town of Hammarstrand was built.

There are hints here of a much more dynamic relationship between humans and rivers – and the deposited sediment which forms floodplains, deltas and other parts of landscapes – than usually accounted for. Flowing water is more than just a passive resource under human control, but an actant in its own right – a force or energy that can influence, resist and act back on human projects. The attempt to elucidate more of the character of that relationship will be picked up again and again throughout the course of this book.

The dark matter of landscapes

Rivers are part of landscapes. Flows of water, and the associated flows of solid materials carried by rivers, are prime movers of landscape change. Considerable amounts of sediment get shifted from one place to another, at once transforming landscapes through erosion and creating landscapes through deposition, often both at the same time (though not normally so rapidly as in the case outlined above). It makes a real difference whether we see people as part of this process or not.

It might be thought, then, that landscape studies and landscape archaeology in particular would be good places to look for the theoretical underpinnings of archaeology of flow. But many books which provide broad overviews of landscape in archaeology and human geography (Muir 2000; Johnson 2006; Hicks et al. 2009; Wylie 2007) make only the barest mention of rivers. There is a clear sense, at least in the British tradition of writing landscape archaeology, in which rivers are seen as something quite different and separate from solid landforms –

even when running through the middle of densely farmed and occupied floodplains or right through the heart of urban townscapes. Patterns of fields and woods, roads and hedgerows, earthworks and village settlements, moated enclosures and manorial estates, buildings and streets, are the kinds of things that are the focus of attention. Rivers are the 'dark matter' of archaeological landscapes.

Part of the reason for this has already been touched upon. Landscape is acknowledged, at least in part, as a cultural construct: rivers are left out of the picture to some extent because they are thought of as natural rather than cultural entities, as liquid rather than solid, and as wet rather than dry. The bottom line is that rivers are not counted as 'land'. But the resulting absence of water flow from the theoretical discourse of landscape archaeology leads inevitably to a somewhat static vision of landscape, with much of its flowing energy and flux conceptually removed.

The place of water in landscape archaeology has up to now been occupied mainly by the subfield of 'wetland archaeology' (Coles and Lawson 1987; Coles and Coles 1989; see also the various articles in *Journal of Wetland Archaeology*). Wetlands include salt-marshes, fenland, peat bogs, river-margins, lake-margins, and so on. A main focus of work has been on the preservation of organic remains and environmental evidence in anaerobic waterlogged conditions, often under threat from drainage, with many startling discoveries of trackways and bog bodies. Almost by definition it is about areas of waterlogged land where there was little flow, such as the Somerset Levels in England and the former bogs in Ireland, as well as wetlands on river floodplains (see Lillie and Ellis 2006 for accounts from wetlands in many other parts of the world).

The book *Rethinking Wetland Archaeology* by Robert Van de Noort and Aidan O'Sullivan (2006), and an article by Francis Pryor (2007) both separately argue that the subfield has become isolated from main currents of method and theory, and make powerful arguments for its realignment back into the mainstream of archaeology. Whether the metaphor they use is intentional or not, the overriding sense they convey is of an area of research that, rather like an oxbow lake, has got cut off from the main current of ideas. There is in fact very little consideration of flow in most archaeological accounts of wetlands (some exceptional examples are discussed below), apart from the flow of people along paths and trackways. Wetlands are clearly important parts of landscapes, but emphasis on near-stagnant pools and bogs should be counterbalanced by a greater concern with flowing water. Extending the scope of wetland research to cover flowing as well as standing water could rejuvenate and revitalise the field of wetland archaeology.

A good example of archaeology of flow within wetland research, however, is the investigation into material structures of the European beaver by Bryony Coles (2006). Archaeological techniques were employed to record dams, lodges, ponds and channels constructed by beavers in their extensive modification of present day streams. As Coles demonstrates, all these structures are best considered not as isolated features but rather as interconnected parts of dynamic systems, contributing and responding to the changing flow and shifting channels of the river. The beaver was a common animal on northern European rivers from the end of the last ice age through to the medieval period, and its constructions were woven into the entanglement of human, river and animal activities over that period.

Emerging evidence and changing perspectives

Some of the best studies in wetland archaeology, broadly defined, are those that attempt to reconstruct dynamic, flowing riverscapes from evidence retrieved in former river courses or palaeochannels, often buried under sediments. Archaeological work carried out near a confluence of the River Trent at Hemington in Leicestershire (Cooper 2003; Ripper and Cooper 2009) is a case in point. Following the earlier pioneering fieldwork of amateur researcher Chris Salisbury in alerting investigators to the great archaeological potential of disused and buried river channels, a programme of watching brief and targeted excavation (especially adapted to the difficult conditions encountered in gravel quarries) was put in place by University of Leicester Archaeological Services. The remains of numerous timber and masonry structures, mostly of medieval date, were uncovered from beneath several metres of gravel. These include a sequence of three medieval bridges along with fish-traps, dams or weirs, evidence of mills, and other structures for consolidating river channels – an array of evidence which testifies not only to flowing water but also to a highly unstable landscape of shifting river channels that people sought to manage and control. Discoveries at Hemington exemplify the theme of the entanglement of nature and culture, and we will return to discuss the site in some detail later in the book.

There are useful guides to the geomorphology of floodplains for archaeologists (Brown 1997; Howard and Macklin 1999; Howard et al. 2003), but it is important to escape from the widespread assumption that, while river geomorphology influences human settlement and other activity, there are no arrows of influence pointing in the other direction. In considering cultural archaeology of floodplains, Brown (1997) and Rhodes (2007) both go some way towards overturning the prevalent environmental determinism and finding a common language between disciplines that generally use very different vocabularies, methods of investigation and modes of argument.

The medieval period in Britain is an area of research where, thanks in part to input of geomorphologists like Brown and Rhodes, the extent of human-river interaction is beginning to be realised, with much emerging evidence and corresponding changes in perspective over the last decade or so. A recent volume edited by John Blair (2007) pulls together evidence for the construction of navigable channels in the late Saxon and medieval periods, looking mainly at the period from 950 to 1200 AD. While it focuses on artificial watercourses for boat transport, it recognises that in many cases navigation was a secondary function, with some 'canals' originally dug for other purposes (mill-leats, water provision or drainage channels, and so on) or as a response to other river structures (bypass channels, for example, were often a response to weirs or other river obstructions). A kind of internal dynamic to the relationship between people and rivers is identified, with the cutting of new channels in part a reaction to the blocking of old ones by weirs and mill dams (Blair 2007: 11). The book brings to the forefront of medieval archaeology a whole new cluster of formerly overlooked linear features of great size, all of which channelled flowing water and entailed modification of rivers in one way or another.

Two recent books on bridges add to this picture. Alan Cooper (2006) and David Harrison (2004) place bridges within broader contexts of road and trade networks and the political economy as a whole, without losing sight of equally important relationships between bridges and river flow and wider environmental considerations. Cooper explains how the general transition from ford to bridge in the tenth to twelfth centuries was brought about not only by the demands of road traffic, but also more indirectly by less obvious factors. More intensive agriculture with associated deforestation and better drainage of fields led to greater run-off of groundwater into rivers. The embanking of rivers, the building of mill-leats and mill dams or weirs, and so on, all combined in complex ways to bring about a deepening of rivers and significant loss of usable fording-places (Cooper 2006: 15-18). As well as being material responses to changing river conditions, bridges – in partially blocking flow – inevitably had their own effects on patterns of sedimentation and erosion. Cooper's analysis reminds us that bridges are structures of flow, at once impacting on and impacted by the currents of the river, as well as to changing political and economic circumstances.

All this augments previous work on medieval flow-engineering in the form of construction of fish-ponds (Aston 1988) and ponds for flax or hemp retting. Most such ponds involved diversion of water from a nearby river or stream through a series of channels and sluices, perhaps by means of a dam, so that water in the ponds could be regularly flushed out and returned to the river, along with accumulated sediment or other residues. As such they should be considered integral parts of river systems, rather than separate adjuncts to it.

There is a growing sense from this body of work, taken together, that all of these different kinds of features – fords, bridges, fish-ponds, mills, weirs, river-fisheries, and so on – were interconnected to each other in complex and changing systems of flow that were more than just the sum of their parts. Consider the field of potential effects and relationships of just one river structure. A weir put across a river to provide a head of water for a mill would deepen the water upstream – a factor in the drowning of fords and their replacement by bridges. In presenting an obstacle across the river, it had obvious implications for passage of boats past the weir itself, leading to the construction of bypass-channels or flash-locks: however, it may have made navigation easier upstream where the water was deeper. The weir may also have blocked the path of fish swimming upstream or downstream, presenting problems for fishing: on the other hand, it presented possibilities for installing fish-traps into the structure of the weir itself. In directing flow into mill-leats or other features, weirs were taking it away from the main course of the river, disrupting other potential uses. Like bridges, weirs would have changed patterns of flow to the extent that erosion of banks would have been caused in some places and formation of islands and shoals in others, leading to further river works becoming necessary. The possible networks of effects are vast, and that is just for a single structure.

It is not just weirs that have such multiple effects. All river installations do. In shaping flow, they have an indirect impact on changing river morphology and other river structures. Through flow, each structure influences and is influenced by all other installations both upstream and downstream on that stretch of river. Political and economic implications of this will become clear when we later look at more detailed examples, but clearly there is great potential for dispute between different types of river-user, and the river system could only have worked through much negotiation and compromise, which means that it has to be viewed as a social or cultural system as well as a hydrological one. As Blair put it, 'there was a symbiotic relationship – normally contentious, occasionally creative – between the uses of flowing water for moving boats, for powering mills, and for harvesting fish' (Blair 2007: 9).

There are ramifications here for geomorphological studies, which in providing detailed accounts of natural processes have often failed to show how these interconnect with cultural processes. An article by physical geographer John Lewin (2010), however, summarises the growing body of evidence for human intervention in river systems from medieval Britain, concluding that 'dynamic floodplain landforms were interactively involved with human development during a critical time period in a totality of ways not previously identified' (Lewin 2010: 267). Lewin's paper is important in that it recognises the impact of cultural activity on river systems (as well as vice-versa), and the need for due account to be taken of archaeological and historical evidence as part of geomorphological investigations. It shows that the very different approaches of physical science and cultural studies can be usefully combined in the investigation of rivers.

Summary

The general proposition put forward in this introductory chapter is that rivers should be regarded as dynamic entanglements of nature and culture. If considered purely as natural systems, their cultural dimension gets excluded. If considered as cultural artefacts through and through, their wild aspect is neglected. The two detailed examples given – the mid-Atlantic streams in North America and the Döda Fallet in Sweden – both illustrate that even the most natural-seeming river stretches may have archaeological dimensions to them, interpretation of which is essential to understanding the wider landscape. I went on to argue that those branches of archaeology which take 'land' as their subject (whether the 'landscape' or 'wetland' variety) should encompass dynamic liquid flows – including flows of solid material eroded, carried and deposited by water – within their remit. In this respect, the archaeology of medieval Britain was identified as a vibrant area of research where – thanks in part to collaboration between archaeologists and geomorphologists – the 'dark matter' of landscapes is coming to light.

2

Unravelling the human-river relationship

Introduction

The last chapter set out the idea that rivers are entanglements of nature and culture. They are part wild, part artefact. Strands of human involvement are folded into the convoluted twists and turns of even the most natural-looking stream, changing their flow and form. Most rivers today can be understood as systems of flow which include the human element as part of their dynamic. As such they are susceptible to archaeological as well as geomorphological and hydrological study. The aim of this chapter, then, is to unravel some of those tangled threads of human involvement, looking in particular at the ancient practices of straightening and diversion of rivers. I use examples taken from recent historical periods, partly because these are better documented and mapped than older examples, partly because any archaeological investigation of rivers has to start with recent modifications and work backwards, in order to understand earlier interventions and interactions.

Historical archaeology is one of the fastest growing areas of research in the discipline today. It can be defined as the study of recently documented periods of human history by using material culture and analysis of landscapes as well as texts (Hicks and Beaudry 2006; Gaimster 2009). When using documents or maps, however, the archaeological focus on the material aspect of things is retained. The following account takes us right into the middle of the fluid environment of snaking meanders of the Mississippi depicted on Harold Fisk's map of 1944 (Fig. 1.1 in the previous chapter). We move from the all-seeing disembodied map-reading perspective looking down from above to the very different point of view of a person (or more precisely, a participant) embodied and situated within the dynamic environment of shifting river channels.

The shifting meanders of the Mississippi: a first-hand account

Mark Twain (1835-1910) can tell us more than almost anyone else about the Mississippi. He not only wrote about life on the river, in his novels *Huckleberry Finn* and *Tom Sawyer*; his own life and identity were river-immersed and river-entangled. As a young man in the 1850s he worked on the Mississippi steamboats and became a master river boat pilot. Even his pen name derived from the call of the leadsman at the front of the boat, who tested the changing depth of the river with a knotted or marked line, providing the pilot with a stream of information on which to base decisions from moment to moment. 'Mark Twain!' refers to the two-fathom mark on the line, indicating that the depth had reached 12 feet, usually a sign that it was safe for the shallow draft steamboat to proceed.

Twain's testimony – Samuel Clemens was his real name – is relevant here because he gained first-hand experience of engaging with the dynamic Mississippi River at a crucial time in its history. His descriptions in *Life on the Mississippi* illuminate key themes of this book, including the entanglement of nature and culture in practice. Although the specific geomorphological conditions of the Mississippi at that time are different from those of any other river (even from the same river today), his observations can shed light on the interpretation of archaeological sites elsewhere, on which the material traces of dynamic inter-relationships between people and rivers are encountered.

Up to the nineteenth century, the strong current of the Mississippi prevented most boats from going upstream. Traffic was largely one-way, with barges or rafts coming downstream with the flow from the upper river to New Orleans. Twain himself describes watching, as a young boy sitting on the levees, rafts the size of football fields stacked with acacia boards, with encampments of makeshift huts on them, being manoeuvred down the river by crews of twenty men or more (the deforestation that was taking place upstream would itself have changed the flow of the river, and the load of sediment it carried, through the soil erosion and increased run-off it caused). Barges had to be laboriously poled by hand all the way back upriver around each and every one of the meandering bends. That all changed with the invention of the steamboat, which had enough power to go against the flow. The whole economy of the river changed as a result.

To Twain the river was the most 'eluding and ungraspable object'. He describes just how difficult it was, while learning to be a steamboat pilot, to memorise the shape of the river:

> I would fasten my eyes upon a sharp, wooded point that projected far into the river some miles ahead of me, and go to laboriously photographing its shape upon my brain; and just as I was beginning to succeed to my satisfaction, we would draw up toward it and the exasperating thing would begin to melt away and fold back into the bank! Nothing ever had the same shape when I was coming downstream that it had borne when I went up (Twain 1883: ch. 8).

Twain was far from being a mere spectator here. As riverboat pilot he was actively engaged in navigating round snags and treacherous eddies, seeking deep water and fast currents, looking out always for sandbars and other shallows that could beach the boat. His standpoint was a mobile one. Not only was Twain's point of view always on the move, but so too was the topography of the river,

> whose alluvial banks cave and change constantly, whose snags are always hunting up new quarters, whose sand-

> bars are never at rest, whose channels are forever dodging and shirking (Twain 1883: ch. 10).

One aspect of these river shifts, he noted, was the way the loops of the river meanders got progressively larger, so that:

> in some places if you were to get ashore at one extremity of the horseshoe and walk across the neck, half or three quarters of a mile, you could sit down and rest a couple of hours while your steamer was coming around the long elbow, at a speed of ten miles an hour, to take you aboard again (Twain, 1883: ch. 17).

He further noted how the river would make cut-off channels or chutes across the neck, diverting flow away from the horseshoe bend. Significantly, the process could be speeded up:

> When the river is rising fast, some scoundrel … has only to watch his chance, cut a little gutter across the narrow neck of land some dark night, and turn the water into it, and in a wonderfully short time a miracle has happened: to wit, the whole Mississippi has taken possession of that little ditch (Twain 1883: ch. 17).

This placed the plantation of the 'scoundrel', whose land was formerly situated in the back of beyond, onto the new river bank – multiplying its value. The old watercourse rapidly shoaled up, becoming impassable to steamboats. Formerly valuable plantations, once but no longer on the economic lifeline of the river, now dramatically lost their value. For this reason,

> Watches are kept on those narrow necks, at needful times, and if a man happens to be caught cutting a ditch across them, the chances are all against his ever having another opportunity to cut a ditch (Twain 1883: ch. 17).

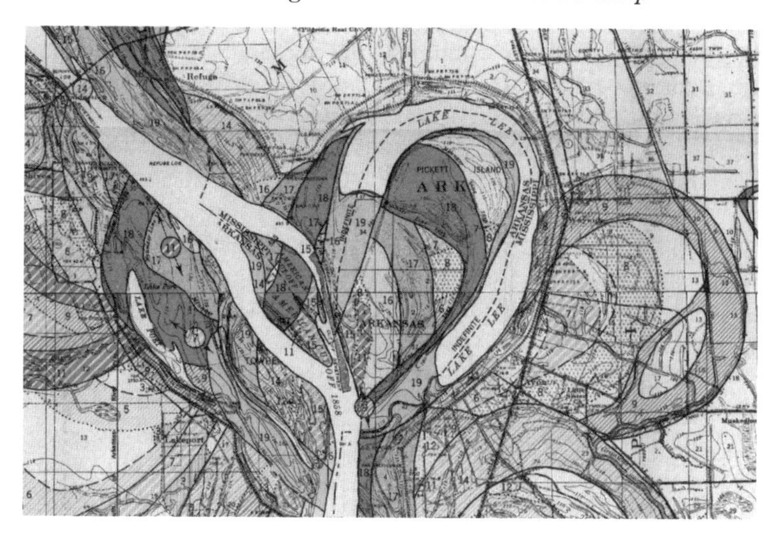

Fig. 2.1. Meander cut-off, artificially cut in the 1850s. Detail of map by Harold Fisk (source: United States Army Corps of Engineers).

In other words, material investment in land along the present course of the river meant that, though some tried to change its course for profit, there were others who would go to considerable lengths to keep it where it was. There was a lot at stake in these battles to retain or change the course, not least for the riverboats, whose journey times could be greatly shortened. An important consideration here too is that the river served as state boundary as well as navigation channel. The cutting-off of a river meander could result in a town in the State of Mississippi suddenly finding itself in the State of Louisiana, or vice versa. Other strange inversions of landscape topology could occur:

> The town of Delta used to be three miles below Vicksburg: a recent cut-off has radically changed the position, and Delta is now *two miles above* Vicksburg (Twain 1883: ch. 1, his italics).

Thus Twain hints at political and economic dimensions to changing river morphology, and human complicity in it, while at the same giving a wonderful phenomenological account of the dynamic state of the river as it was then. He was *there*. He was caught up in it all. He was in direct contact with the river (through the steamboat and the leadsman's line). At the same time he portrays the Mississippi as more than just a river – as a dynamic assemblage of forces and flows and materials and artefacts all interacting with each other. One of the flows is the traffic in both directions of steamboats themselves, together with the passengers and the cargo they carry. Cut-offs or chutes can be the work of rivers, or people, or an entangled mixture of both. Not only are nature and culture intertwined, but so too are materials and ideas. Economic motives of greed and profit work both with and against the flow of the river.

The Civil War brought an end to the steamboat era. Years later Twain returned to see a very different Mississippi, its flow more controlled, its length shortened, its meanders curtailed, its levees further extended and built up on either side. But like Wild Huss, the logging merchant on the River Indal, he retained something of the wildness of the river in his character (for rivers shape people as well as the other way round). To him, a river pilot in the heyday of steamboats was 'the only unfettered and entirely independent human being that lived in the earth' (ch. 14). Later in his career, something of that came through in his writing.

Turnbull's Bend

The act of digging a semi-artificial chute in order to cut off the loop of a meander might seem insignificant, relative to the construction of other archaeological features and monuments. But here is an example of how the unintended consequences of such an act can be enormous. In the early nineteenth century one of the many great loops in the Mississippi, forcing steamboats to travel thirty miles or more out of their way, was called Turnbull's Bend.

Fig. 2.2 shows the sequence of river development that led to formation and transformation of this river feature. Red River

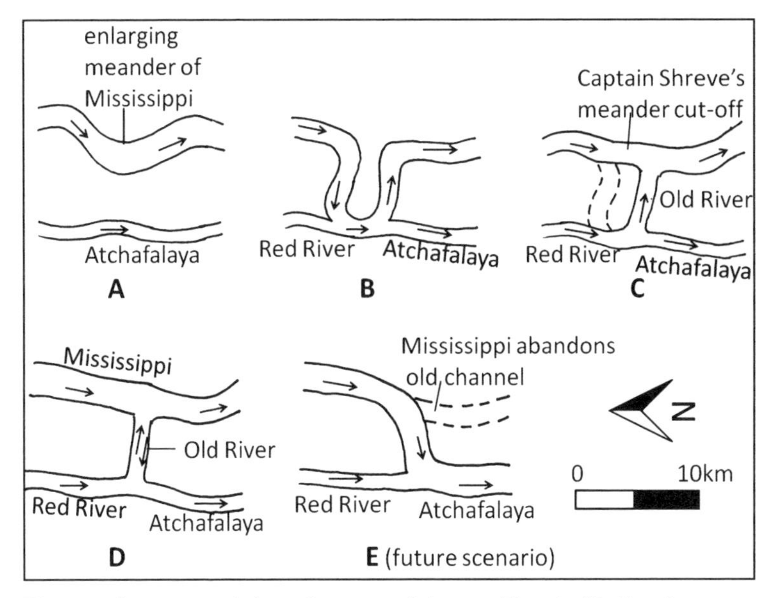

Fig. 2.2. Sequence of river flow transitions at Turnbull's Bend.

and the Atchafalaya River had once been a single watercourse separate from the Mississippi (A). It was intercepted by an enlarging meander of the Mississippi sometime between the eleventh and fifteenth centuries, turning the Red River into a tributary, flowing in, and the Atchafalaya into a distributary, flowing out (B). In 1831, a steamboat captain called Henry Shreve took matters into his own hands and ordered the digging of a cut-off channel (C). All it took was the cutting of a narrow channel and the power of the river did the rest. The cut-off quickly widened to become the main course of the Mississippi, allowing the northern part of Turnbull's Bend to silt up, while the southern part, now known as Old River, continued to flow (Clark 1983: 83-5).

Not much water flowed into the Atchafalaya because part of its course was blocked by a huge logjam. Horseriders could cross it without knowing there was a river underneath. According to McPhee (1989: 40), it was so jammed that herds of cattle could be driven across the logs, where river and cattle-trail intersected. But in 1839 the State of Louisiana blasted and dredged the blockage away to turn the Atchafalaya into a navigable channel, increasing its flow. This had the unintended side-effect of presenting the Mississippi with a shorter and steeper route to the sea. The direction of flow along Old River now reversed (D) – or at least, it flowed east when the Red River was high and the Mississippi low, and west into the Atchafalaya when the Red River was low and the Mississippi high. The Atchafalaya started to draw more and more water from the Mississippi (10% by 1890, 30% by the mid-twentieth century). The more water it took the broader its channel became, taking even more water. By the 1950s it had became clear that a major river capture was about to take place. The very real prospect of the Mississippi changing its course completely (E) was averted only by massive river engineering works.

It does not fall within the remit of this book to describe in detail the complex system of dams, new channels, locks, levees and other devices (comprising the so-called River Control Structure) that then had to be built at Turnbull's Bend by the Army Corps of Engineers (see Clark 1983: 83-7). Suffice to say that without these measures, the Mississippi mouth would be about seventy miles to the west of where it is now (McPhee 1989: 42-92).

The Mississippi almost certainly *will* move into the Atchafalaya channel eventually. Most hydraulic engineers agree on that. It is just a matter of how long human ingenuity can keep the inevitable at bay. The consequences of such a shift would be enormous. Towns like Morgan City on the Atchafalaya are likely to be initially inundated but might benefit economically in the long term. That part of the Louisiana coastline would be reshaped. Meanwhile the flow of the former course of the lower Mississippi River would be reduced to a tiny proportion of what it is now. The present Mississippi delta, already sinking through loss of river-borne sediment (another consequence of river control), would be massively eroded. Industries that depend on the river would go into decline. Cities like New Orleans and Baton Rouge would become swampy backwaters, losing their status as deepwater ports, leading to huge population movements away from these areas and towards the new delta.

The important point to emerge from all this is that apparently small human actions can play a large part in bringing about great changes on river systems and landscapes. The intervention that Captain Shreve made in digging a shortcut in order to circumvent the great loop of Turnbull's Bend, and the subsequent act by the State of Louisiana in blasting a navigable channel through the logjam on the Atchafalaya, are no different in principle or scale from the kind of human engagement with rivers that was going on throughout medieval Europe, or at other places and times to be discussed in this book, except in being better documented. Studying such recent events can help us to grasp something of the character of human-river interactions in general.

To intervene in river processes is to get entangled with the river, because it invariably responds to the intervention in multiple ways that are not entirely predictable, requiring further interventions. In this case actions were successful in bringing about intended outcomes (the opening of shipping channels), but there were also totally unforeseen effects. No one anticipated that the combined effects of the two actions (along with other factors, such as increased velocity and volume of flow brought about by deforestation upriver) would be to open up the possibility of the Mississippi being captured by the Atchafalaya. Subsequent interventions were made to meet that threat, but these in turn have had their own unpredictable effects, requiring further responses, and so it goes on. Entanglement in this sense, though a partly metaphorical term, really does describe something of the character of the relationship that develops between people and rivers.

Rivers in the city: the city in rivers

Rivers are entangled in urban development too. It is customary to say that a town or city has been built 'on' a river, with the implication that the river is part of the environmental background and that the city is overlaid on top, rather as a cultural layer might be overlaid over the top of a natural layer in a sequence of archaeological stratigraphy. Actually it would be more accurate to say that in most such cases the city and the river are *interleaved* with each other.

Consider a typical pattern often encountered in the archaeology of medieval towns. Riverside urban areas were used for mooring boats, with quayside structures installed. Flooding occurred from time to time which covered the quayside with silts. Rubbish was dumped into the river from urban markets, shops and houses nearby, pushing the bank further out into the river. Partly as a flood prevention measure, the river was embanked to keep its flow within the banks, but the river now overflowed its banks more often because of the increasing narrowness of its channel. The growing town expanded over the old quayside area right up to the embankment, with more rubbish dumped as landfill so that another quayside area could be built on top. More flooding and dumping occurred, which led to further embankment. Thus the process went on, as the town gradually pushed out into the river.

This process of encroachment leaves its own characteristic signature in the archaeological record, found in excavations of medieval waterfronts in London, Dorestad, Bergen, and many other river towns and cities. A series of timber or stone revetments set one in front of the other are discovered, with the earliest further away and the latest towards the river, interleaved with layers of alluvial silt laid down by the river and midden-like dumps of landfill material (Milne 2002; Milne and Hobley 1981).

In such sequences and patterns of stratigraphy, at least, the river and the city are truly intertwined. Flows of solid and liquid materials are mixed up together. Thus at Reading a huge dump of clay revetted with timber was initially used to create the waterfront area in the twelfth century. This was extended further in the fourteenth century with the construction of a silt trap. A wattle fence trapped particles of silt while letting water through – the accumulation of sediment then being revetted with heavier posts and planks to form a substantial quayside (Muir and Muir 1986: 90-3; Hawkes et al. 1997).

Smaller streams may be just as enmeshed in urban development. Birmingham Archaeology recently conducted excavations on numerous sites within the previously neglected historic core of the city where flowing water was a vital force, so crucial for urban generation. What these investigations revealed was the important role played by fast-flowing small streams in proto-industrial development in late medieval times (Patrick and Rátkai 2008). Antiquaries John Leland and William Camden, travelling through Birmingham in the sixteenth century, remarked upon the extraordinary vitality of the place – 'echoing with forges' – as they ascended the main road of Digbeth from watery Deritend to the more fashionable part of the town atop the sandstone ridge. It was this steep gradient that was part of the key to Birmingham's success as an early industrial centre. Small streams that poured down the slope from a spring-line were used to service forges and fill tanning pools (as well as to run through moats and boundary ditches). In transformations of matter from one state to another, the flowing energy of water was harnessed to power bellows to fan the furnace fires that melted ores and fashioned solid artefacts out of molten metal, then utilised in the cooling processes too. Metalworking and cloth-processing were specialities. The principal mills for grinding grain for the town were on the River Rea, with lesser mills sited on smaller streams used for sharpening tools or weapons, wire-drawing, metal-rolling, and so on. Where a bridge and ford took the road across the River Rea, a causeway doubled as a dam and served to divert the river to run alongside the road for a short distance, its flow being diverted through a network of channels, tanks and pools and its energy turned to numerous industrial tasks. At least in Birmingham, water flow was an important part of the material conditions which made the Industrial Revolution possible, even if its role was later forgotten.

Many people, of course, may be surprised to hear that Bir-

Fig. 2.3. Where is the river in this picture? The Westbourne River runs through the metal culvert above the platform of Sloane Square Station, London (photo: sunil060902, 2008, CC by 3.0).

mingham even has a river. The River Rea now runs mostly in deep open culverts below the level of the city streets, out of view of motorists passing by, though at least still under sky. That is not the case, however, for the so-called 'lost rivers' of London (Barton 1992), many stretches of which are completely enclosed, encased and entombed – integrated into the city sewage system. Few train passengers are aware, as they stand on the platform at Sloane Square tube station in London, that the huge iron pipe crossing the line above their heads is actually the formerly vibrant and sparkling Westbourne River (Barton 1992: 133).

Such streams – not just the Westbourne but the Fleet, the Walbrook, the Tyburn, the Effra, and more – were once vital to the growth of communities that later coalesced into the great industrial city. Along the riverbanks were osier beds and fisheries, basketmakers and rushcutters, tanning yards and breweries, mills and forges. It is only in the last two hundred years or so that they have been forgotten. But even now – encased in metal or brickwork, buried under concrete, filled with slurry and undermined by railway tunnels, woven like pieces of thread into the material fabric of the city – there is still a sense in which such rivers possess a wildness that cannot be entirely contained within culturally applied forms. The main branch of the stream might be entombed, but a river also consists of the tributaries that flow into it from all parts of its watershed, and many of these still run underground along the course of former streams. On reaching the original beds but unable to join the main flow such influxes of water are forced to run through the water-bearing strata alongside, *outside* the culverts, causing much trouble for builders and railway engineers (Barton 1992: 134).

LIBRARY, UNIVERSITY OF CHESTER

The Rhône at Lyon: moving the confluence

Modifications to rivers can be hidden by later development. Take a city like Lyon in eastern France, on the confluence of the Rhône and Saône rivers. Walking round or studying a map of the city can be puzzling for anyone trying to work out the sequence of urban expansions from Roman times to the present day. To make sense of the Lyon cityscape it is essential to be aware of one extraordinary historical event. In the late eighteenth century, a monumental river diversion took place. *The whole river confluence was moved by about a mile downstream.*

Knowledge of this radical re-shaping of the rivers in Lyon is key to understanding the urban layout of Lyon as a whole, and its trajectories of development today. Only then can we grasp, for example, how the present day Confluence Project – perhaps the largest urban renewal project to take place in Europe since the post-war years, adding 370 acres of prime real estate to the historic city centre and effectively doubling it in size – is possible. The moving of the confluence and reclaiming of land over two hundred years ago is historically recorded, but rarely made much of in narratives of city development. It is referred to in the World Heritage listing (UNESCO 1998) only as drainage work. I happen to know of it only through family history as a relative was involved in the work. I draw from his memoirs (Edgeworth 1820) in recounting part of the little-known story.

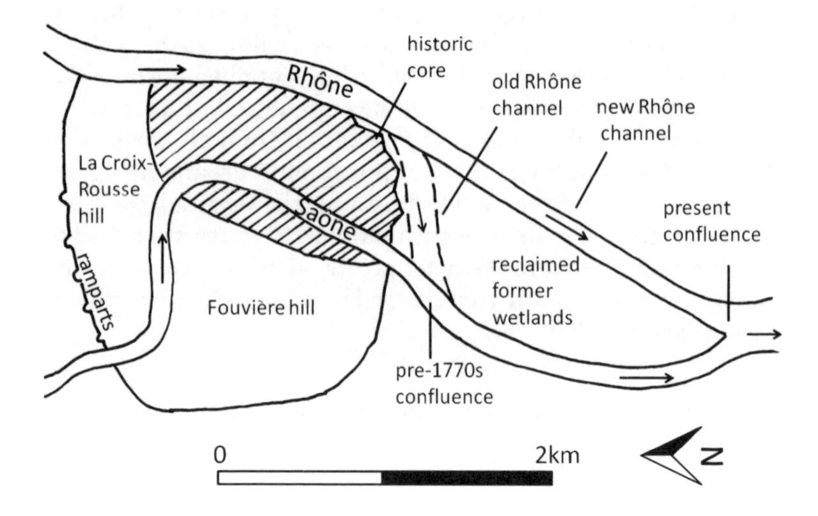

Fig. 2.4. Map of Lyon showing former and altered locations of the confluence.

Plans were first drawn up to move the confluence, in order to expand the city on its southern side, in the 1720s. It was not until 1771, however, that sufficient funds were raised to form a company. Work started in 1772, led by the architect and sculptor Michel-Antoine Perrache. He employed a young Irish engineer called Richard Lovell Edgeworth (RLE), who was put in charge of operations. The aim was to turn the Rhône from its course and divert it into a new channel, to join up with its tributary the Saône further downstream, thus allowing the reclamation of land between. The new channel only needed to be partially cut: the flow of the river itself, once diverted, would excavate most of its own bed and carry away the spoil. The bed of the old channel would be filled with city rubbish and provide yet more land to build upon.

Nearly a hundred men were employed to construct a dam of piles and earth across the Rhône. The flow of the river was deployed in this process too. An ingenious kind of ferry, common on the Rhône, was used to transport material across the

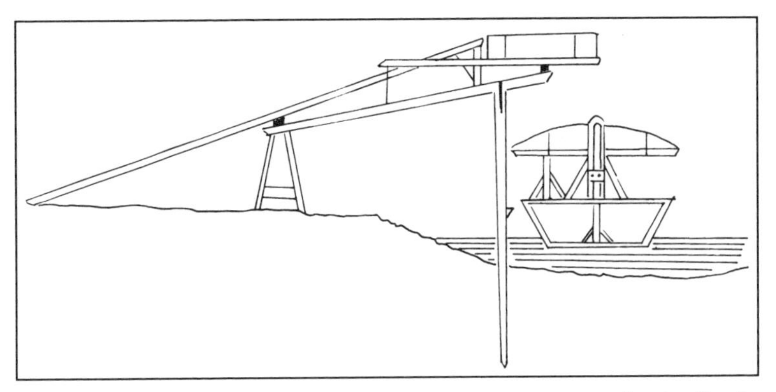

Fig. 2.5. Machinery deployed at Lyon, 1772, swept away by the flood.

river to the far side of the dam. The ferry was attached by ropes to a cable suspended from high poles on either bank. When set at an oblique angle in relation to the current (acting in effect as its own rudder) it could travel in either direction across the river, using no other force than the flow of water.

Dam construction proceeded well until a gap of about twenty feet remained to be closed. The full force of the river rushed through this gap. Dumped material was being carried away by the current faster than it could be deposited. The problem was getting men with wheelbarrows to transport earth quickly enough, and valuable time was wasted waiting for the ferry to complete its journey. This prompted the design of a series of time-saving devices, including a moving platform at the top of a ramp that could be filled with earth while the ferry was away, then rotated on an axle to release its load at once on the boat's return. A similar moving platform was fitted to the boat, so that the entire contents could be emptied into the river in one go, exactly in the location desired.

A further innovation was an early kind of conveyor belt. On a trestle bridge – built across the excavated channel that was to be the new course of the river – a number of wheelbarrows were mounted on a moving cradle drawn by pulleys. Crowds of people gathered on the city ramparts to witness the extraordinary sight of a succession of barrows, filled with earth, apparently wheeling themselves across the bridge, without men to guide them.

The work on the Rhône diversion was nearly completed when a boatman warned of an impending flood. RLE implored the company to employ more workers and to increase wages so that work could be finished before the flood reached Lyon. Not understanding the risk that flood posed, the company was unwilling to find the necessary money. This is how RLE described what happened next:

> At five or six o'clock one morning I was awakened by a prodigious noise on the ramparts under my window. I sprung out of bed. I saw numbers of people rushing towards the Rhone. I foreboded the disaster! ... When I reached the Rhone, I beheld a tremendous sight! All the work of several weeks, carried on daily by nearly a hundred men, had been swept away. Piles, timbers, tools, and large parts of expensive machinery, were all carried down the torrent, and thrown in broken pieces upon the bank (Edgeworth 1820: 316-17).

Although the river diversion was eventually accomplished, it took another fifty years for the reclaimed land to be properly drained. It was used first as the site of a mill and city dump, with much burning of rubbish to dry it out. In the mid-nineteenth century a gasworks was built, followed by a huge railway terminal. Then came slaughterhouses, prisons, brothels and wholesale markets, with the banks of the Saône developed as docks. It was an industrial, liminal zone – the dark side of the city, downstream of its cultural heart. Something of the river still hung over it, like a river fog rising from the damp ground.

That all changed with the ambitious Confluence Project, initiated in 1998 and costing over a billion euros. Now the industrial installations have been demolished and the land is being redeveloped as a residential and business zone, with tourist attractions and prestigious public buildings. These new developments are only possible as a result of the work carried out before, marking the latest phase of an ongoing historical transformation of the river city.

Rivers and developing technology

There was a certain amount of scientific hubris involved in the moving of the confluence in Lyon – the belief, current in the late eighteenth century, that nature could be 'conquered' by science. The elaborate machinery and the spectacle it provided for an admiring populace were all a part of that. The relationship between science and the river was not one of conquest, however, but rather one of entanglement. It was a two-way material transaction. By engaging with the swift current of the Rhône, by trying to find solutions to the problems it presented – literally pitting wits against it – the full force of Age of Enlightenment technology was brought to bear upon the river and to some extent transformed by it. The river provided material feedback which sparked a degree of creativity and inventiveness in designs for further interventions. Machines were adapted to the flow of water, just as the flow of water was itself shaped and modified.

Far greater engineers than RLE were greatly influenced by their engagements with flowing water – not least Leonardo da Vinci (1452-1519). While working as a military engineer for the city-state of Florence, in collaboration with Niccolò Machiavelli, he attempted to divert the River Arno away from the city of Pisa, which was under siege. In 1504, thousands of labourers were set to work in building a dam across the Arno upstream of Pisa, and work was also started on two channels which would serve as new courses of the river. Sophisticated designs for excavating machines with which to speed up digging were put forward. The project failed, however. When the river flooded it was deflected by the dam into the diversion channels as planned, but as the floods subsided the water reverted back to its former channel and partly destroyed the dam – probably because Leonardo's instructions as to the depth of the digging had not been properly carried out. As soon as the works were abandoned, forces came out from Pisa to remove all traces of the dam and fill in the diversion channels (Masters 1999: 93-133). The example shows how rivers can get tangled up not just with the development of technology but also with political and military power struggles. This is not the only instance in this book of rivers and their flow being used as weapons of war.

From an early age Leonardo had been fascinated by patterns of flow, as his maps of the River Arno and his many drawings of eddies, vortices, currents, turbulences and deluges indicate (Ball 2009a: 1-12). His encounters with flow clearly impacted on his art, for many of his paintings of various subjects seem to have the tremendous vitality of surging water (see Rubens' copy of his lost painting, *Battle of Anghiari*). His scientific innovations were also greatly influenced by flow, which was in turn shaped by his inventions. In particular, Leonardo's version of the pound-lock – with hinged V-shaped mitre gates, self-sealing when closed by water pressure upstream – became an important element in the much later industrial revolution of the eighteenth century. 'Staircases' of such locks made possible the interconnection of rivers and canals into a network of waterways, facilitating transport of goods and materials across large distances.

Flowing water in one form or another is a participant in what Bruno Latour calls 'socio-technical collectives'. To Latour and other actor-network theorists, it makes no sense to try and separate out the artificial from the natural, the human from the non-human, because these are inextricably mixed up together. According to his principle of symmetry, one should not prioritise one side of any duality over the other, but always look for their mixture or entanglement (Latour 1993; see also Witmore 2007, Witmore and Webmoor 2008 for elaboration into a symmetrical archaeology approach). The principle applies not only to modern industrial artefacts but to the study of human technology and material culture in general, because it is always a mixture or assemblage of natural and cultural forces, human and non-human materials and agencies. Rivers should be seen as part of such assemblages.

It was of course the transformation of water into steam that powered the industrial revolution of the late eighteenth and nineteenth centuries (Mithen 2010: 5250), and before that it was the energy of rivers exploited by the waterwheel that powered much of the equally significant but often forgotten industrial revolution of the medieval period (Gimpel 1992). Industrial zones were either placed next to rivers – often downstream of towns where pollution could be taken away by flow – or rivers were brought to industry through the creation of artificial distributary networks similar to those described in Chapter 4. Potteries, breweries, tanneries, ironworks, dyeing works and textile mills all used flowing water as a material resource and a source of energy. Both Birmingham (with its metal-working) and Lyon (with its silk-working) owed their growth as industrial centres in large part to their rivers.

Political and economic aspects of the urban river: the Garonne in Toulouse

The Garonne which flows through Toulouse, like the Rhône in Lyon, is a fast-flowing river. Both have sources in high mountains – the Rhône in the Swiss Alps and the Garonne in the Spanish Pyrenees – deriving much of their flow from melting snow. Both were susceptible to Spring floods, with the capacity to damage or destroy structures placed across the river like bridges and dams. In Toulouse, the strong current of the Garonne was extensively utilised for powering watermills. In the early twelfth century, there were over sixty boat-mills, clustered in three groups on the river (see Gräf 2006 for a detailed account of so-called 'floating mills').

A problem with boat-mills, however, is that they tended to break their moorings and be swept out of control, crashing into other water installations. In Toulouse there were so many of these floating mills that the problems were compounded, and the millwrights collectively decided to replace them with a series of stone weirs serving fixed mills on the river bank. Two excellent accounts of the development of the medieval industrial structures on the river at Toulouse will be drawn from here. The first, by Gimpel (1992), looks at the relationships between the weirs and political organisation of the mills. The second, by Fortune (1988), focuses on the relationship between the weirs (and bridges) and aspects of river morphology. Taken together, they give a holistic account of a dynamic river system that combines political and social with material and hydrological factors, always in relation to the direction of flow of the river.

Weirs were built at Château-Narbonnais at the upstream entrance to the town, La Daurade near the centre, and Le Bazacle at the downstream end. The falls of water thus contrived powered numerous mills on the right bank (facing downstream). The Bazacle weir, built on the site of a former ford, was almost a quarter of a mile long, placed diagonally across the river to offset the force of the current. It was built by ramming two rows of massive oak piles into the riverbed, then filling the spaces between with stones and gravel. Breakwaters were constructed just upstream to further protect it from floodwaters and the debris carried with it. The earliest mention of the weir is 1177 (Gimpel 1992: 17-18).

Gimpel draws attention to a crucial interrelationship that bound the structures together into a dynamic though unstable system. That is, the fall of water on any one weir was always an effect, not only of its own height, but also of the height of water retained by the next weir downstream. The Bazacle weir, being the furthest downstream, was the only one that could set its own height independently of the other weirs (Gimpel 1992: 18).

There were not just hydrological interconnections, but political and economic ones too. Gimpel tells of the succession of lawsuits and counter-suits in the thirteenth and fourteenth centuries as owners of weirs unlawfully heightened them in order to get a better fall of water. Each time one weir was heightened it meant that the weir upstream had to be heightened too, which impacted on the next weir, and so on. The middle weir and its mills, unsurprisingly perhaps, were put out of business in 1408 (Gimpel 1992: 18-20).

In a classic analysis, Gimpel puts forward the case for the Société du Bazacle being the oldest limited company in the world. From the thirteenth century until it was nationalised in the mid-twentieth century, 'shares of mills on the Garonne were subject to annual fluctuations in value' and 'bought and sold freely like shares on a contemporary stock exchange', yielding annual dividends. There were annual general meetings where managers were elected. Because there were no millers among the shareholders, who were mainly wealthy citizens of Toulouse, there was a clear 'division between capital and labour' (Gimpel 1992: 13-23).

According to Gimpel, it was the sheer complexity of interrelationships – not only between different clusters of weir and mills and organisations associated with them, but also between the various other interests such as fishing and navigation (all of which pertain to flow in one form or another) – that gave rise to the birth of capitalistic enterprises. As in the modern city (see Swyngedouw 2004: 2), the flow of water and the flow of money were materially linked.

In a complementary account of the same stretch of river, Fortune uses documentary evidence to reveal a pattern of frequent and sometimes disastrous flooding of the town, and the damage these inflicted on the river structures. To him, the floods were the direct result of the damming of the river by the weirs. He uses historic map analysis to chart changes in the shape of the river over time, noting that there were once numerous islands in the river, which gradually disappeared over the centuries. There was a corresponding loss of channels, with one arm of the Garonne lost completely through erosion and other channels silting up through deposition of alluvium. A former channel called the Garonette, once used for the outflow from the mills of Château-Narbonnais, was filled in and turned into a street in the 1950s. Weirs caused erosion and sedimentation in equal measure on different parts of the river, and this impacted back on the river structures themselves. Continually damaged by floods, they had to be frequently repaired or rebuilt completely, bringing about further effects on the flow of the river. One rebuilding of the Bazacle weir, for example, led to the formation of large alluvial deposits upstream and an elevation of the level of the river bed, creating the material conditions for further and even more serious floods (Fortune 1988: 183).

As a response to the floods, linear earthen embankments were built to protect many of the islands. This led however to swifter currents between them, countered by some islands being removed, the effect of which was offset by the deposition of further alluvium downstream and the formation of more islets (Fortune 1988: 184). It was a fluid situation, with each intervention bringing about its own assortment of unwanted side-effects, needing further interventions to solve them. The river changed shape as islands fused into each other or separated into smaller islets, the flow of the river changing accordingly, much of it as a direct result of the weirs. Weirs in turn were repaired, replaced, modified or redesigned to take account of changes in the river and the damage it wrought. Eventually in the late nineteenth century the Bazacle weir was remodelled to provide the fall of water for an early hydro-electric power station.

To fully appreciate the story of the Bazacle weir, we have to find a way of integrating Gimpel's analysis with Fortune's account of changing river morphology. Ultimately the two aspects – the socio-economic and the geomorphological – are not separate but inextricably intertwined as part of the same unfolding set of material relations. Archaeology of flow, as conceptualised in this book, is a place where these different aspects can be studied together.

Summary

This chapter set out to unravel some of the entanglements of nature and culture in rivers. As soon as we started to look closely at any river of recent times, however, the more it became apparent that separating out the river narrative into separate strands of natural and cultural is impossible. Hydrological, social, geomorphological, political, material, ideational, symbolic, technological and economic factors are all mixed up together, like so many streamlines merging into a single current. What emerges is a dynamic of human action and river response, a kind of multiple feedback system that unfolds through time in not entirely predictable ways. Once caught up in that dynamic, even the smallest of material interventions, like the digging of a chute on a Mississippi bend back in the 1830s, or the placing of a weir across the fast flowing Garonne, can lead on to consequences out of proportion to the original act. In the last example of the Bazacle weir in particular, we saw that the fluid entanglement of nature and culture can have effects far beyond the river itself, reaching into the heart of administrative and political systems, perhaps even helping to kick start (if Gimpel is right) the capitalist enterprise itself. This makes structures of flow such as meander cut-offs and weirs – up to now so neglected as to almost not exist as classes of archaeological feature in their own right – much more important than previously thought.

3

Traces of flow: the material evidence

Introduction

River diversions and partly artificial meander cut-offs are not just modern phenomena. The last chapter took up the perspectives of historical archaeology to briefly examine some well-documented examples of human-river entanglements from relatively recent times – enabling us to apprehend something of the character of interactions between people and flowing water. In this chapter we look further back in time. Although the specifics of cultural, historical and geographical context are very different in each case, some basic elements of the relationship are broadly similar. For example, the resistance of the river encountered at Lyon, with floods destroying months of work, would be all too familiar to river engineers diverting rivers in much more ancient periods. Interactive effects of one river structure on others both upstream and downstream were not at all unique to the weirs at Toulouse; this is something that has to be dealt with on all intensively modified stretches of river, ancient or modern. And the struggle or wrestle with the river between competing groups of people – some of whom were actively taking steps to alter its course, some of whom had a vested interest in retaining the existing shape of the river – was not just a feature of life on the Mississippi in the nineteenth century. Evidence for all these can be found at much older sites, for which documents do not always survive and where archaeological investigation of material remains may be the only means of bringing human interaction with rivers to light.

The River Trent at Hemington: a dynamic confluence

The River Trent in the English Midlands is a dynamic, unstable river, and nowhere more so than at its confluences with the Rivers Derwent and Soar near Hemington. But the sinuous movements of its meandering channels over the last thousand years or so cannot be regarded as separate from human activity. The construction of fish-weirs, fish-traps, mill-dams, roads, bridges and other structures was interwoven with the changes of course in the shifting loops of the river (Clay and Salisbury 1990; Cooper 2003; Ripper and Cooper 2009).

The sheer range of structures found from under several metres of alluvial and fluvial deposits is extraordinary. Perhaps the most spectacular are the remains of a sequence of three medieval bridges. All the bridges were relatively short-lived and used different construction techniques. The first, dating from the late eleventh to the late twelfth century, was a trestle bridge built on lozenge-shaped timber caissons filled with sandstone blocks to make pier bases. The second, slightly further upstream, lasted from the late twelfth to the mid-thirteenth century and employed the more conservative method of pile-driven posts. The third, built to replace the second but abandoned in the early fourteenth century, had masonry piers supporting a timber walkway (Ripper and Cooper 2009).

Bridges present both obstacles and openings to water flow, often creating vortices; these can cause scouring and undermining of piers on both upstream and downstream sides. In an unusual and groundbreaking collaboration, insights gained from archaeological excavation and detailed analysis of sediments were combined with those of hydrology, sedimentology and geomorphology to show how, in the case of the earliest bridge, scouring from severe flooding was the principal cause of its destruction. One of the caissons had been laterally rotated and another tipped from the horizontal, with the trestle falling on top of it – all these structural components settling into deep scour holes, some over 2 metres deep relative to the riverbed.

Fig. 3.1. The earliest of three medieval bridges at Hemington (photo: University of Leicester Archaeological Services).

There was also clear evidence, in the form of scattered masonry and associated deep scour holes, that flood destroyed the third bridge too (Brown 2009).

The different techniques used for each of the bridges shows that attempts were being made to adapt to difficult river conditions. But meanwhile the river was changing as well, and part of this was adaptation to the structures built within the river.

Other major structures discovered were two large dams or weirs, dated to the twelfth century, running diagonally across relict channels. These were constructed of parallel rows of posts about 8 feet wide, the space between them filled with large stones. Their principal function was to create the head of water necessary to drive mills, diverting flow of water through sluices into millraces (Clay and Salisbury 1990; Cooper 2003: 36-7). One of the weirs took further advantage of the controlled flow by incorporating a fish-trap into its structure, of which more will be said below.

Also recorded in the quarry were numerous V-shaped fish-weirs dating from the Middle-Saxon to the medieval period (Cooper 2003: 32-6). Lines of posts close together, joined by

Fig. 3.2. Stone weir running diagonally across old river channel at Hemington. The fish-trap can be seen in the lower right hand corner. Direction of flow is from top right to lower left (photo: University of Leicester Archaeological Services).

closely-woven wattle hurdling, formed the lines of the V. Fish-baskets would have been attached to form the apex, creating in effect a giant funnel in which fish would be trapped. It makes a real difference which way the V (and basket) is pointing. If upstream, then the trap would have been intended to catch fish swimming in an upstream direction against the flow (such as salmon migrating upstream to spawn). If downstream, then the trap was intended to catch fish swimming downriver with the flow (such as silver eels on their migratory route to the Sargasso Sea – see discussion below). At Hemington there are examples only of the latter, although both kinds are represented elsewhere on the Trent.

Exploitation of river resources was one thing, but keeping

the flow of a dynamic and unstable river under control was another (Cooper 2003: 38-9). Several examples of bank-side works of stone jutting out into the stream, apparently designed to deflect current and prevent bank erosion, were found downstream of the stone dam or weir – presumably to prevent further meandering of the river, or perhaps to avoid the formation of meander cut-offs. A significant probability is that these were protecting at least partly against erosional effects of the weir itself.

Paul Courtney's detailed analysis of the wider landscape through historical research provides crucial background for the archaeological evidence at Hemington – placing it within the broader context of the road network, and the flows of people and animals and goods that crossed the river at that point. Courtney also reconstructs something of the political context of local estates and toll territories (Courtney 2009: 174-206). When it is considered that the river itself forms a boundary between estates, it can be easily understood why great effort might be put into retaining (or attempting to change) the form of the river, especially with all the material investment of bridges and other works. A change in river course – or indeed in the position of the river crossing – could represent a massive disruption in the political as well as the material landscape, to the benefit of some stakeholders and to the detriment of others.

The stretch of river at Hemington was truly a 'taskscape' (Ingold 1993; Van de Noort and O'Sullivan 2006: 60), where flow was one of the main elements of the landscape that was being fashioned and used. The material structures and associated sediments all testify to a wrestle with flow – people interacting in complex ways with a dynamic and changing river. Such discoveries show what might be found in other floodplain contexts in Britain, but tend to get missed by standard evaluation methods, such as trial trenching and geophysical survey. Strategies adapted to meet the demands of quarry conditions at Hemington, combining expert watching brief with targeted excavation, could serve as a model for future investigations in gravel quarries elsewhere.

Artefacts of flow

The well-preserved wicker fish-basket found in association with the stone weir or dam at Hemington is an artefact of flow in more ways than one. First of all, it derives something of its vortex-like form from flow, having been adapted and 'streamlined' to flow conditions. Second, it has been constructed in such a way, and of such materials, to funnel flowing water while at the same time allowing it to pass through its woven materials. Third, it was affixed to a solid structure (in this case a mill dam or weir) which was itself contrived to control and channel flow in particular ways. Fourth, in order to work efficiently it had to be oriented properly in relation to the direction of flowing water. Fifth, as well as being oriented in relation to river current, it may also have been oriented in this case in relation to a much greater flow – the movement of eels in their migration downriver on the way to their spawning grounds in the Sargasso Sea.

The trap was in the form of a cone-shaped basket over 2 metres long, containing two smaller internal funnels which acted as non-return valves for any fish that went through (Cooper 2003: 37). Fish-traps of similar design are known from other rivers in Britain, and indeed from ethnographic contexts all over the world.

The fact that the basket trap was placed with its entrance facing upstream and its pointed end downstream indicates that fish swimming with the current were the intended catch. In the middle of summer mature silver eels leave backwaters and ponds of river valleys to swim downstream on the initial stage of their four thousand mile journey to the ocean to spawn, and it is likely that the trap was set specifically to intercept this migratory flow (Tesch 2003; Cooper 2003: 37-8). It was the life-force of the migrating eels themselves – the urge to spawn – that impelled the fish into such traps.

There are of course many other kinds of artefacts of flow. Without the space to discuss them in this short book, it is worth making some general points. Artefacts of flow can be said to

Fig. 3.3. Wicker fish-trap from mill weir structure at Hemington (photo: University of Leicester Archaeological Services).

have one thing in common: a fundamental orientation in their design, affordances and operational capacities to flowing water, as well as to human agency and to networks of other artefacts. Because of that orientation, it is only by placing them within interpretive frames of reference which take flow into account (as part of the dynamic past assemblages of human and non-human materials and agencies in the context of which such objects were used), with interpretation itself oriented around the basic directions of upstream/downstream, that archaeologists can make much sense of them.

Modes of investigation and analysis

At Hemington, survival of archaeological structures and artefacts was facilitated by the fact that the river suddenly changed course during a flood, burying its former channels under deposits of gravel. Paradoxically, the evidence was preserved by violent river movements, while the gentle meandering movements of less volatile rivers might be more destructive in eroding traces of human-river interaction (Howard et al. 2008). Some rivers are relatively stable with little channel movement, and an important question is how archaeological techniques can be applied to flowing rivers, as opposed to buried palaeochannels. Work of Louis Bonnamour (2000) on the Saône in France and Attila Tóth (2006) on the Drava and Danube Rivers in Hungary is of interest here. Tóth discusses the pros and cons of bathymetric survey, sonar and seismic radar in mapping the riverbed, and describes how different kinds of objects and structures – from log-boats to dams and buried walls – have been located through such methods. He also points out that underwater exploration by diving can be extremely useful, especially when combined with other methods (Tóth 2006: 64). Underwater archaeology is by no means restricted to sea and lake, although rivers do present especially challenging conditions, with particular problems of visibility. Timber piles of bridge foundations or stone debris from collapsed bridges are examples of structures found by divers. Objects and structures on the riverbed identified through sonar can be further targeted by underwater survey and excavation.

Artefacts retrieved from rivers – whether obtained through diving on the riverbed, metal-detecting and other prospection on tidal foreshores, or from examination of mud taken out by river dredgers – can provide important archaeological information. A problem here is lack of precise context, and the likelihood that river currents have moved discovered objects from their original points of deposition. Artefacts could easily have found their way into the river through erosion of riverside cemeteries or settlements. Even if deposited directly into the river, artefacts may have been carried considerable distances by the current – sometimes grading the artefacts in the process and depositing all those of a particular size in the same places, giving a false impression of finds concentrations. Yet assemblages of artefacts from rivers are still valuable sources of evidence, if carefully interpreted with effects of flow in mind. An example is the Thames Water Collection at Reading Museum, which contains over 500 objects found by dredger crews from non-tidal parts of the River Thames from 1911 to 1980. Finds range from Mesolithic flints to post-medieval artefacts, and include many Bronze Age and Iron Age weapons.

Rivers often demarcate boundaries between administrative and political areas, but at the same time may also run through the centre of other territories, serving to unify as well as divide, and contributing to collective senses of regional identity. Charles Phythian-Adams argued for the existence of what he called 'cultural provinces' in England, corresponding roughly to the watersheds of rivers (Phythian-Adams 1993: 1-23). So although rivers might sometimes be liminal entities, running along the edge of inhabited zones, they also play a central role in people's lives and consciousness, as environments rich in resources and as arteries of trade and communication flowing through the heart of landscapes. Rivers can of course be part of sacred geographies too. Christopher Woods gives an intriguing account of the symbolic relationship between locations of sacred cities in early Mesopotamia and the geomorphology of the Euphrates floodplain. He equates the duality of river-gods – often represented on clay tablets as paired dyads, male and female – with opposite banks of the river (although other river features such as confluences and bifurcations could be relevant here too). Beliefs about the Euphrates as the 'creatrix of everything' were enmeshed in settlement patterns and systems of irrigation often conceived of by Wittfogel and others in purely materialistic terms. Woods argues, for example, that the veneration of river gods at cities like Sippar and Hit – sited at either end of a distinct and recognisable geomorphological zone – was to some extent a function of river geography. Hit was located at the 'true beginning of the alluvium – at the point where the Euphrates emerges from its deeply cut valley, thereby making gravity-flow irrigation, and thus life itself, possible' (Woods 2005: 40). Sippar, on the other hand, was located at the point where the river started to fan out into the lower alluvium. According to Woods, then, worship of river deities at both these locations was bound up with a practical knowledge of floodplain morphology.

Rivers are multi-scalar entities that can be studied at many different scales of analysis (see Ball 2009b: 104-11 on the fractal structure of branching river networks). Google Earth and other GIS applications have transformed the study of rivers, allowing investigators to look at flowing watercourses all over the world, zooming in and out to view at different scales at the touch of a button. Far more than just a collection of aerial photos, such software makes use of digital terrain data collected from NASA's Shuttle Radar Topography mission, allowing 3D viewing and analysis of watersheds. Computer visualisation and modelling are increasingly useful tools for studying river development, with much interpretation and prospection for new sites taking place on screen as well as out in the field. Whole new types and classes of evidence, the existence of which was previously unsuspected, are coming to light as a result.

Former river installations can sometimes be spotted submerged underwater, either through sighting of the structure itself – such as the 260 metre long V-shaped medieval fish-trap recently discovered in the Teifi estuary, visible from over a kilometre up (the reader can use Google Earth to find this structure for themselves). Disturbances in patterns of flow can also indicate the presence of unseen structures below the waterline. Another possibility is identification of river structures from variations in vegetation growing on the riverbed – in effect underwater 'cropmarks' (especially if the aerial photo was taken when the water was low and clear). An example of this will be mentioned in the next chapter.

The book *Where Rivers Meet* by Simon Buteux and Henry Chapman provides a good example of the use of advanced computer techniques combined with traditional fieldwork methods such as excavation, land survey, augur survey, geophysical survey and aerial photo analysis. Taking the Trent-Tame confluence in England as subject area, the book examines a complex of sites in the vicinity of the river dating from the early prehistoric period to 900 AD. Like so many confluences, this one was the focus of much ritual activity and monument building during the Late Neolithic and Early Bronze Age (Buteux and Chapman 2009).

A puzzling feature of the evidence presented in *Where Rivers Meet*, however, is that there is no indication there of past human engagement with flowing water or actual modification of changing river morphology. In fact, there seems to be absence of any hint of the dynamic interplay between humans and rivers that we have noted elsewhere. In that sense the findings of the Trent-Tame confluence study contrast with results already discussed from the Trent-Derwent-Soar confluence at Hemington a relatively short distance downstream, where numerous river structures from the medieval period indicated human intervention in river flow and form, together with river response and the countermeasures provoked. In this earlier period of study at Cathome, however, rivers are portrayed as though they provide merely the environmental background for cultural activity, rather than being part of an entanglement between natural and cultural forces.

A question that arises is whether prehistoric periods in England were really characterised by such lack of engagement with flow, or whether apparent lack of evidence for human-river entanglement comes about for other reasons (poor survival of evidence, different techniques of investigation, etc.). It is well known that the medieval period saw a great increase in use of water power for milling and corresponding development in river technologies. The same may be true to a rather lesser extent for the Roman period. But does that mean that there was little physical interaction with rivers in Britain in earlier times? Physical geographer Roger Lewin, having comprehensively reviewed evidence of medieval entanglements with rivers, notes that 'two-way interactions between human enterprise and environmental change are also likely to be true of long-past periods of development' (Lewin 2010: 268). Is there any evidence, though, to back that up?

Natural places?

It would be strange if British prehistory was indeed devoid of evidence of such two-way interaction, given the focus of much prehistoric settlement along river valleys. Rivers and their floodplains obviously provided rich resources in terms of fertile soil for farming and plentiful game for hunting and fishing, as well as corridors for movement of people and goods through the landscape. That rivers were also rich in symbolic meanings is evident from the many examples of ritual monuments clustering around river sources and confluences, as Buteux and Chapman themselves show at Catholme. Kenneth Brophy notes the close connection between cursus monuments and rivers; most lie on floodplains or river-terraces and several cross or are crossed by streams. Indeed, he argues they are symbolic rivers (Brophy 1999). It seems highly likely then that there must have been associated material intervention in the form and flow of the rivers themselves.

Going by general discussion of rivers in archaeological theory for the prehistoric period, however, it would seem that the main evidence of human-river interaction is provided by votive deposition, especially around crossing-places and springs. We have already touched upon this in the discussion about artefacts retrieved through river dredging. A general pattern of votive deposition in wet places is noted by Richard Bradley (1990) and other writers (Van de Noort and O'Sullivan 2006: 58). Bradley argues that rivers have a particular association with deposition of weapons, although he acknowledges that different stretches of the same river can have associations with different kinds of artefacts (Bradley 2000: 54, 56). Julian Thomas says that rivers and other wet places often received distinctive sets of deposits in the late Neolithic, including whole Peterborough ware vessels and unbroken stone maceheads (Thomas 1996: 177). Riverine contexts of votive deposition generally seem to be thought of as natural places and sacred places at the same time. As Bradley explains:

> Natural places have an archaeology because they acquired a significance in the *minds* of people in the past. That did not necessarily make any impact on their outward appearance, but one way of recognising the importance of these locations is through the evidence of human activity that is discovered there (Bradley 2000: 35).

There is a definite sense here in which flowing water is regarded by scholars as part of the natural background *onto* which cultural meaning was applied or *into* which cultural items were placed, without having any cultural dimension in its own right. There must come a point, however, at which so-called 'unaltered places' are significantly transformed by sheer volume of objects placed on the riverbed. In one 20 kilometre stretch of the River Ljubljanica in Slovenia, for example, over ten thousand artefacts ranging from Neolithic antler implements to Iron Age spearheads and swords to complete sixteenth-century pottery vessels have been found, and that represents only a tiny proportion of what must have been deposited in the past, as part of a tradition of river deposition which seems to have lasted for thousands of years (Kaufmann 2007).

The reason why the River Ljubljanica was so culturally significant may be to do with the fact that it runs through extraordinary karst land formations riddled with underground cave networks, so that the clear fast-flowing waters of the river suddenly disappear into or well up out of the ground. In Britain it is well-known that many monuments cluster around river sources and springs. The phenomenon of 'emergent flow' (flowing water emerging from caves or springing up from the ground itself, as at many holy wells) was clearly a focus of veneration and ritual in prehistory and later times.

It seems likely that many artefacts were deposited in rivers as part of actual physical engagements with flowing water, often in conjunction with river structures which rarely survive in the archaeological record. Most appear as discrete items only because their material contexts of deposition have been swept away. But there are exceptions. Bronze Age spearheads were found alongside a jetty of similar date at Vauxhall on the River Thames in the 1990s (Haughey 1999). At Testwood in Hampshire important discoveries were made by Wessex Archaeology in advance of the creation of a new reservoir. Three Bronze Age bridges over relict stream channels were found, with a bronze rapier and part of a Bronze Age boat found in association with one of them. A worked bone object sharpened to a point was found in association with late Iron Age riverbank revetments (Wessex Archaeology 2008).

Bridges and revetments are both to some extent responses to flow, while at the same time having impact on currents and sedimentation patterns. As material interventions into river morphology, they would have prompted river responses which required further interventions, along the lines of the dynamic human-river interactions already examined.

The view of rivers as natural places, then, implies that past meanings were somehow detached from material practices which might have impacted on flow. I think this is wrong, and that there may be a wealth of evidence yet to be discovered of human-flow interaction in British prehistory. This applies not just to rivers but to many sites where channels of flowing water were incorporated into designs and layouts.

Symbolism does not necessarily imply a lack of engagement with rivers. River crossings are almost inherently given to symbolic treatment, and may be culturally embellished with a host of beliefs and rites, perhaps involving ritual immersion in water or votive deposition of artefacts (possibly as gifts to river deities) but often manifested through the physical act of wading through a ford or crossing a bridge (Edgeworth and Christie 2011). Rites of crossing physical boundaries like rivers may be synonymous with rites of passage that help mediate transition across social boundaries (Van Gennep 1908).

Other kinds of boundaries such as pit alignments could be related to rivers in ways not so far considered. Digging ditches across meander loops was a technique deployed in other times and places to deliberately straighten the course of rivers. Here the intention may have been precisely the opposite. It is known that many pit alignments have a close association with rivers, often cutting off meander loops or following the course of disused river channels (see Buteux and Chapman 2009: 108); if these boundaries had been dug in the form of ditches they might have altered the course of the river itself (in the manner of the artificial chutes discussed in Chapter 2). It is possible that some boundary markers were deliberately dug in this alternative form, using spoil from pits rather than ditches to construct linear banks, as a strategy not to impact upon or change existing patterns of flow (which as part of a sacred geography they may have wanted to preserve). Such conscious non-involvement with rivers does however imply an awareness and knowledge of changing river morphology derived from other entanglements with water flow.

Symbols are all the more powerful when embodied in actual material practices. Consider the story of the royal burial of Gilgamesh, as told in the *Epic of Gilgamesh* (Foster 2001), which takes us back to the beginnings of state formation in early Mesopotamia. Following cryptic instructions received in a dream, the king of Uruk ordered workmen to construct a dam across the Euphrates River, and to dig a new channel into which the flow could be diverted. A tomb was constructed on the riverbed into which offerings to the gods could be placed. After Gilgamesh drank poison and lay down in the tomb, the dam was breached and the river diverted back again, sealing the burial with the flowing waters. In this legend at least – written down on Assyrian clay tablets of the seventh century BC and referring to events in the mid-third millennium BC – there is a physical entangling of culture and nature (not just a conceptual one). The material and symbolic aspects of the river are woven together, the symbolic thread looped beneath that of the river itself. Although there is no evidence that riverbed burials were carried out in British prehistory, it is likely that there was considerable enmeshment of material and symbolic aspects of rivers.

Tracing the human-river relationship back in time: dams

The extensive damming of rivers for hydro-electric power, irrigation, industrial supply, etc., that has taken place since the late nineteenth century – transforming major rivers into staircases of vast linked reservoirs – is often held to be the main process through which some rivers were changed from natural to culturally-controlled entities. Many rivers today, like the Colorado River in the USA or the Yellow River in China, have had so much of their flow intercepted and diverted that they sometimes fail to reach the sea. Effects on water distribution and sediment deposition of vast structures like the Three Gorges Dam on the Yangtse River in China or the Aswan Dam on the River Nile in Egypt are cited in arguments for the existence of the Anthropocene – a new geological epoch marked by human impact on earth systems and cycles, measurable in geological terms (Zalasiewicz et al. 2008).

The accelerating extent of river transformation in modern times is undeniable. Even so, it would be wrong to assume that before the start of the Industrial Revolution (often taken to be the start of the Anthropocene) rivers were largely unmodified by human action. We have already seen – in the example of the mid-Atlantic streams discussed in Chapter 1 – how easily it can happen that material traces of past human-river entanglements get disguised by later transformations, which might appear to us to be natural. It was noted that mill dams on rivers in parts of Europe in the medieval period and earlier probably had a comparable impact on deposition of sediment and floodplain morphology.

The importance of dams and weirs in medieval cities has already been touched upon. In creating a head of water for mills, they provided power for industry. In configuring urban riverscapes into upper and lower levels, and sometimes into series of stepped levels like descending staircases, they structured all subsequent human-river interactions on that stretch of river. In diverting flow into subsidiary channels, they made

Fig. 3.4. The Harbaqa dam between Damascus and Palmyra in Syria (photo: Armin Hermann, 2006, CC by 3.0).

possible the many examples of artificial distributary networks to be discussed in the next chapter. In the Netherlands, both Amsterdam and Rotterdam are named after their dams. In England, Warwick, Ware and Wareham were named after their weirs.

Romans employed most basic dam designs, including gravity dams, arch dams, buttress dams, barrage dams and embankment dams. The dam at Harbaqa in Syria is over 20 metres high and 365 metres long. Note how it has completely silted up with sediment on one side, the disused dam being used now as a road (Fig. 3.4). In the case of Harbaqa the build-up of sediment was an unintentional effect of the dam, whose main purpose was water-retention, but the principle that dams could be used to manipulate flows of more than just water was intentionally applied elsewhere. In arid regions of the Middle East, Near East and North Africa, dams were constructed across wadis with the specific aim of capturing the rich and fertile silt that flash floods carried. With soil easily eroded, alluvial and colluvial sediment built up quickly behind dams, effectively turning the valley into a stepped series of fields, and the dams into terrace walls. Similar techniques were then applied to terrace the valley sides, providing large surface areas of cultivatable soil for farmers to work. Vita-Finzi describes soil terraces built up behind such dams in post-Roman North Africa as typically 2-4 metres high: 'this deposit, largely silt, incorporates fragments of weathered limestone and also Roman pottery from the hilltop farms, and fragments of dam masonry' (Vita-Finzi 1960: 63). In his detailed study of desert landscapes of Roman Arabia, Graeme Barker traces such practices back to the Early Bronze Age, with increasing sophistication in techniques of manipulating flowing water (and flowing soil) over the next few thousand years (Barker 2002: 497-9). One such check dam in the Near East has been dated to 3955-3630 BC (Wilkinson 2003: 190), and there could be much older examples.

Each of these ancient dams on its own would of course be overshadowed by the modern Aswan Dam, but the sum total of all the smaller dams and the sediment they retained in many wadis across large areas and over long periods of time actually amounts to a major landscape transformation of similar magnitude, leaving a huge physical trace in the archaeological (and geological) record.

Larger early dams include the Sadd el Kafara dam in Wadi Garawi, Egypt. It is 110 metres long and 14 metres high, with a base width of 98 metres and a crest width of 56 metres. It is thought to date from the mid-third millennium BC. An earth and rubble core was consolidated by rockfill either side, faced with limestone ashlars. The exact function of the dam is not known with certainty, but its location halfway between the River Nile and a large alabaster quarry has led to the suggestion that it was designed to provide material (both water and sediment) for watered paths or mud roads needed for the transport of large stones on sledges – literally *stone flows*, as illustrated on papyri (Fahlbusch 2009). However it seems almost certain that this and other early surviving dams, such as the Jawa Dam in Jordan – often said to be the oldest dam on earth – must have had precursors of earlier date. Herodutus tells of an Egyptian legend in which the River Nile itself was diverted by the construction of a dam, so that the river ran along the east side of the valley rather than the west, thereby ensuring enough space for the founding of the city of Memphis (Jansen 1980: 1). Such legends, like the story of the royal burial of Gilgamesh, are unverified and need to be approached critically – but even so provide tantalising hints of early human-river interaction on larger scales than might otherwise be imagined.

There is considerable archaeological evidence of fish-weirs dating from the Mesolithic period in northern Europe, America, and elsewhere – such as those recently uncovered 6.5 metres down in the mud of the Liffey estuary, Dublin (McQuade and O'Donnell 2007). It might be supposed that such structures were built on inland rivers too. Fishing by means of weirs and traps was widespread among hunters and gatherers and cultivators across the world long before the colonial period. Remarkably similar types of fixed weirs and baskets encountered in widely different times and places are a testament to the extent to which these structures were adapted to flow, deriving their form partly from the characteristics of flowing water itself. These were by no means purely functional installations, however. Some such structures can be tied in with complex systems of ritual beliefs and practices (Losey 2010).

Ethnographic accounts of fish-weirs abound from many parts of the world. The Brewarrina fish-traps on the Barwon River (a tributary of the Darling River) in Western New South Wales, Australia, are usually assumed to be many thousands of years old. These complex arrangements of stone walls in the river marked places where groups of Aboriginal people gathered for trade and rituals as well as fishing. According to native tradition, the traps were founded by the ancestor Baiame and his two sons Booma-ooma-nowi and Ghinda-inda-mui. Stories of their journeys and adventures are woven into the physical structure of the traps, which are part of the landscape of the Dreamtime (Dargin 1976).

Extensive networks of low linear earthworks and associated ponds are found throughout the Amazonian region of Baures in Bolivia. Covering hundreds of square kilometres, the earthworks were recently shown by Clark Erickson to be fish-weir structures, which took advantage of seasonal flooding to capture fish. The low banks were typically set out in zig-zag formation, with funnel-like openings for channelling fish from the retreating floodwaters into artificial ponds (Erickson 2000). Dating from the pre-conquest period, and probably thousands of years old, what is especially interesting about these structures is that they have emerged through detailed survey from an area previously thought to have been largely unmodified by human activity. This reinforces the recurring theme of this book – the extent to which supposed 'natural' environments can hide ancient structures of flow that played a significant part in transforming and creating landscapes.

Anthropologist Hugh Raffles was surprised when, during ethnographic fieldwork in the Amazon basin in Peru, a local Indian casually mentioned that the nearby river had been cut by hand. This led to his detailed exploration of the anthropogenic modification of rivers in the book *In Amazonia* (Raffles 2002). A parallel account of the digging of a meander cut-off on the Ucalayi River (Abizaid 2005) is reminiscent of Mark Twain's description of similar activities on the Mississippi, showing that people using simple tools can play a major role in transforming rivers and floodplains. Such work challenges the widely held assumption that the physical environment of the Amazon River and its tributaries is largely unmodified by human intervention. Amazonian people, contrary to widespread popular and scholarly perception, do 'intervene in fluvial systems, manipulating rivers and streams to modify the landscape' (Raffles 2003: 165).

How old, then, is the human entanglement with rivers? All the examples given in this book have been of fully modern humans, within the Holocene period. But river entanglements probably go back many hundreds of thousands of years before that into our hominid ancestry, and their close association with rivers and flowing water. Although studies of technology in human evolution tend to focus on solid artefacts like hand-axes, it is likely that the cultural transformation of the material domain – and corresponding cognitive development – also took place in relation to fluid entities and the manipulation of flow. From the remote times when hominids placed stones across a stream to step across to the other side, built a crude dam with sticks in order to create a pool for fishing or bathing, or modified a beaver dam to suit their own purposes, they were starting to influence and control the flow of water, bringing about river responses which would have required further intervention. Entanglement with flowing water in rivers may extend right back into deepest prehistory, and be interwoven with the course of human evolution in ways not yet explored.

Channelling flow with levees

It is useful to discuss levees here because they illustrate more clearly the mixture of nature and culture than any other kind of river structure, and also the connection between past and present human interventions at any given moment in time. Running more or less parallel to direction of current these linear earthworks help to channel flowing water along a given course. They may be designed to stop lateral river movement, deepen the channel, prevent flooding, and so on. Like dams, they have effects on flow both upstream and downstream as well as at the point of intervention. I use the example of the Yellow River in China.

The Yellow River is over 2,800 miles long, rising as a clear stream in the Tibetan uplands. It cuts steeply for 600 miles through a plateau of loess, or wind-blown silt, originating from the Gobi Desert and the Siberian tundra. Here it erodes and picks up vast quantities of the powdery soil – giving the water its characteristic yellow colour – so that on leaving the plateau

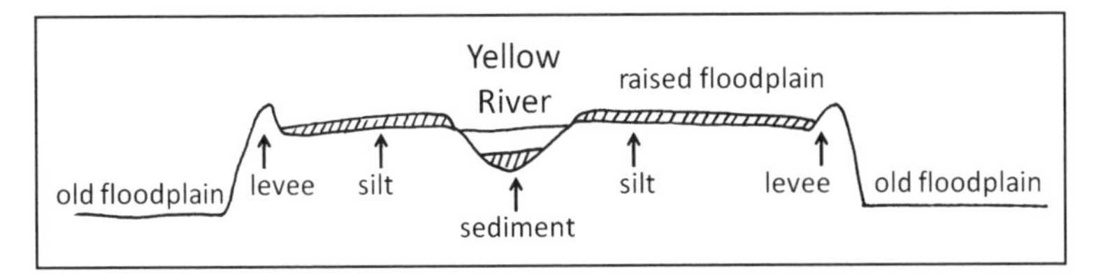

Fig. 3.5. Schematic profile across the Yellow River, its levees and raised floodplain (height exaggerated).

up to 45% of the weight of the water is actually silt held in suspension. It is not just a flow of liquids, then, but solid and liquid mixed up together – a kind of 'muddy fluid' or 'fluid mud' even more viscous than the Mississippi.

When the river descends from the plateau onto the North China Plain, the gradient lessens and the current slackens, resulting in deposition of sediment on the bed of the river, which raises it up. As the river floods it also deposits sediments along the sides of the river, forming natural levees. The more the riverbed rises above the surrounding floodplain the more people need to protect themselves from flooding by building up the levees on either side. That is where the Catch-22 kicks in. The augmented levees prevent the river from depositing its load over its floodplain, and the only place for the sediment to go is on the bed of the river itself, raising the level of the river yet further. This makes it necessary to heighten the levees even more. And so the vicious cycle goes on. The riverbed gets elevated higher and higher, in some cases up to 10 metres above roads and houses and fields on either side, hence it is sometimes called the 'hanging river'.

The levees are immense earthworks, constructed mainly of redeposited river silt. They divide flowing water from land, but unlike fixed boundaries are often in a state of flux, moving in both lateral and vertical directions through time. They have been built and repaired over hundreds of years by countless labourers using (until recently) only the simplest of tools. Note however that the elevation of the river by means of levees cannot simply be described as artificial. It is a complex interleaving of natural and cultural forces, with transport/deposition/erosion of sediment being carried out by both river and people. Today the full weight of modern technology in the form of giant scoops, suction pumps and diesel-powered earthmoving machines is enlisted into this dynamic assemblage of forces, flows and vibrant materials.

An old principle of Chinese river modification encapsulates the ideal of natural and cultural forces intermingled and working together: 'A good canal is scoured by its own water: a good embankment is consolidated by the sediment brought against it' (quoted in Dodgen 2001: 14). But the levees do not quite exemplify this principle and the Yellow River carries too much silt in it for such a state of equilibrium to last long. In other words the silt is deposited on the riverbed faster than the flow of water can scour it away. Meanders inevitably start to form, even within the constraints imposed by the levees. These cut into the levees on the outside of the bends. Further levees are built midstream to contain the flow that causes the erosion. As sediment inexorably raises the river bed, seasonal floods threaten to break through. Eventually a flood occurs so powerful it breaches or overtops the levee.

While most breaches are repaired quickly before the river escapes completely, sometimes this is not the case. In 1887, the Yellow River burst through a breach which quickly widened to half a mile wide, spreading waters and silts in a vast lake of 10,000 square miles across the floodplain, drowning villages and enveloping the landscape under a carpet of silt up to 6 feet thick, but leaving its riverbed dry. A massive labour force was quickly mobilised to plug the gap. A correspondent at the time described the scene witnessed from the top of the bank as thousands of labourers worked with spades and wheelbarrows, shifting material from the abandoned river bed onto the levee:

> We look down some 40 feet on the river below, which pours through a strait about 400 feet wide, with a current of eight or nine miles an hour, in a stream 100 feet deep.

> Huge whirlpools in the centre of the gap show the immense force of the volume of water ... What now are the materials with which it is proposed to force this body of water, much against its will, into its channel? They are five: sticks, stones, stalks, sand, and bricks (*North China Herald*, October 1888, quoted in Clark 1983: 47).

This is a scene that has occurred many times in the history of the Yellow River, going back thousands of years. Importantly, a breach was not always the result of mere flooding alone; sometimes political and social factors also came into play. When for example the Japanese army advanced on a key railroad junction in 1938, Chinese forces enlisted the powerful force of the river itself as ally. They cut an artificial breach in the levee so that the river poured through the gap towards the invaders. Whatever short-term military gain was obtained, the long-term effects were disastrous. The entire Yellow River escaped from its channel to flood over 9,000 square miles of plain as it sought other routes to the sea. Hundreds of thousands of people were drowned and millions made homeless. The plain was still flooded when the Japanese surrendered in 1945. Only after the war in 1946 were engineers able to return the river to its former course.

This kind of human intervention on the Yellow River is far from unique. Neither is it a modern phenomenon. Similar episodes occurred in a dispute between warring dynasties in 923, and again in a peasant revolt of 1642. In the latter case the levees were breached by rebels in a deliberate move to flood the city of Kaifeng, which was under siege. Over 300,000 people are thought to have died in one of the worst 'natural' disasters ever, and the city was temporarily abandoned (Lorge 2005: 147). Given that an effective method of protecting against flood on one side of the river would be to breach levees on the other side, there have been many times in the past when levees needed to be defended against human as well as water erosion.

Major floods of the Yellow River leave massive amounts of sediment in their wake – enough to bury cities. The city of

Kaifeng is extraordinary in that the greater part of the outer city has been buried by flood sediment and rebuilt on top many times. Between the level of modern buildings and the buried remains of the medieval city of Dongjing or Bianjing, for instance, lies an immense 8m of silt. Underneath the ruins of this earlier city lies a further 2 metres of sediment, which covers the ancient buried city of Daliang, dating from the Warring States period of the fifth to third century BC (Johnson 1995). Daliang in turn lies above yet more sediment, and there may be further archaeological horizons below that. Lower layers were once much thicker, but were compacted down through the sheer weight of the accumulated river silt and cities above. The capacity of the river to move immense amounts of material has enormous implications for the preservation of archaeological evidence, and many other sites must await discovery (see Lawler 2010 for an account of recent work on a buried Han Dynasty village at Sanjangzhuang, close to another former course of the Yellow River). It would be difficult to find stratigraphic sequences that show more emphatically the intermingling of 'nature' and 'culture'.

Just 9 kilometres north of Kaifeng runs the present course of the 'hanging river' – where the river channel is raised by levees 10 metres above the surrounding floodplain. Part natural features and part ancient monuments (though yet to appear on any World Heritage Listing), the levees are earthworks that the people of the floodplain have strong interests in maintaining. Should the levees on the south side of the river be breached again (and it seems inevitable that this will happen sooner or later), a catastrophic disaster of major proportions would result. As far as the archaeological record is concerned, further inundation of water and silt would add to the stratigraphic sequence described above.

Sometimes the escaped river cannot be made to return to its channel. This happened in 1861. After a six-year attempt to repair the levee breach, the river settled into its present channel, emptying into the sea about 500 miles to the north of its old mouth. The former river channel was left to dry out and erode, with great implications for settlement patterns and movements of populations.

There have been at least twenty-six major changes in the course of the Yellow River over the last 3,000 years. Sometimes it occupies one of its previous channels or flows into another river, effectively taking it over. Or it may carve a completely new channel. The important point is that this cannot be seen as just a natural process: people have played their part in it too. It is all part of the dynamic human-river entanglement. And each time the river changes course the process of levee building and river bed elevation begins all over again.

Documents of the third century BC reveal that the Han Dynasty government took control of levees on the Yellow River, creating a unified system of flood control and irrigation. But the practice of levee-building is much older than that, and probably goes back to the Late Neolithic (third millennium BC), when large numbers of villages occupied the floodplain. Archaeology clearly has much to contribute in investigating the incredible levees of the Yellow River.

Levees epitomise the dual structuration of river structures. Material legacies left behind by past human/river interactions (containing possibilities and constraints for further action) condition the range of potential interventions that can be made, and likewise any interventions made will leave a material legacy for future generations to deal with.

A problem is that archaeological accounts are few and far between. As we have seen, rivers are put into different categories from other cultural artefacts, earthworks or monuments. They are still mainly seen as natural or environmental entities despite the obvious extent of human modification and investment of human labour. In this section I have turned to historians, geographers, geomorphologists and others to try to unravel the human-river entanglement somewhat, drawing in particular from Clark (1983: 37-53), Dodgen (2001), Jiongxin Xu (2003), and Mei-e and Xianmo (1994).

Summary

This chapter has reviewed evidence from sites of archaeology of flow. In Britain, material remains from both Hemington (medieval period) and Testwood Lakes (prehistoric periods) show that old river channels buried under river gravels in mineral extraction quarries can contain structural and artefactual evidence of exceptional preservation, shedding light on past engagements with rivers. Recent developments in computer mapping techniques have enabled remarkable work to be carried out on ancient river landscapes, both in temperate and arid zones. Through examining some of this work, a question mark was put over whether rivers in British prehistory can be framed within the category of 'natural places'. This questioning forms part of the general argument of the book that rivers – even prehistoric ones – should be regarded as entanglements of nature and culture.

Human-river entanglements extend through time as well as through space. Discussion of the levees on the Yellow River showed that chains of human intervention and river responses over time can lead to the gradual construction of monumental rivers – running like elevated highways above streets and rooftops of the cities on the old floodplain below. Some of the historical trajectories of human-river interactions described in this chapter, unfolding over hundreds or even thousands of year, have played a part in shaping contemporary situations. That is one reason why it is important to understand past human involvement with rivers, through the study of their history and archaeology, so that strategies for the future can be based on consideration of developments leading up to the present moment.

4

Flowscapes

Introduction

Flow is a part of landscapes, not a-part from it, and archaeologies of flowing water should therefore be integrated into conventional archaeologies of landforms and solid materials. In most cases the same techniques and methodologies – excavation, documentary research, earthwork survey, site visits, geophysical survey, augur survey, LIDAR, study of maps and aerial photos, and so on – are relevant to both. But there are certain rationales and principles that are particularly applicable to considerations of flow. In thinking about how human action has both shaped and been shaped by water flow in the past, it is useful to start from a very basic but crucial aspect of the behaviour of rivers and streams, relative to land gradients and the pull of gravity. The principle is simply this: *water flows downhill.*

There are of course exceptions to the general rule. In his environmental history of the Columbia River, Richard White describes how Indians were able to take their canoes upstream *with* the flow by staying close to the shore, taking advantage of eddying currents (White 1995: 9). Many aqueducts of classical times employed an inverted siphon technique to take water down and up the sides of valleys (Hodge 1992: 147-60), reminding us that water can flow uphill under pressure. Direction of flow regularly changes on tidal stretches of rivers like the Thames in London, due to the power of the tide countering gravity flow. Tidal bores are waves which travel upriver. There are also examples of flow being reversed on some rivers through engineering works, such as the Chicago River in the nineteenth century.

It is also important to bear in mind that land gradients can change over time as a result of other processes. Thus many former river courses of prehistoric and Roman periods in the fenlands of East Anglia survive today as slight embankments known as 'roddons', due to the shrinkage of peat soils on either side of the old riverbed deposits, so that water runs off them rather than into or along them (Muir and Muir 1986, 53-4).

Despite such qualifications, however, the general rule that water runs downhill has always been of great use to manipulators and shapers of water flow, as indeed it is to those who seek to make sense of material traces of flow in the archaeological record. Working out movements of rivers over time, and reasons why people modified watercourses to their own ends or created artificial layouts incorporating flow, always takes place by reference to gradient. Gradient is something that theories of flow can be tested against. At the very least, the slope of any given terrain or surface must be taken into account. The effect of the pull of gravity on water flow on different surfaces makes some things possible and other things difficult or impossible. It gives scope for human ingenuity but also imposes limits on it. Gravity gives water an energy that people had to either work with or fight against. Because it enabled and constrained what people could do with flow in the past, it also enables and constrains the archaeological interpretation of those activities.

From archaeology of solid materials to archaeology of flow

One of my first encounters with flow in the archaeological record came when I worked as supervisor on the excavation of a subterranean stone structure at Earl's Bu in Orphir on Mainland Orkney, in the late 1980s (Batey and Morris 1992). The structure consisted of what appeared to be a tunnel leading into a chamber, and was assumed at first to be a souterrain (that is, an underground habitation, store, refuge or ritual sanctuary). But as investigation proceeded evidence began to emerge of a very different function. The 'tunnel' that 'entered' the chamber turned out to be too small to admit a human being. The funnel-like form of the 'chamber', partially lined with heavy clay, suggested the movement of water rather than people. And when the idea that the supposed tunnel could in fact be a tail-race of a mill began to take shape (originally the suggestion of local farmers, who were familiar with similar systems of flow still in use as drainage tunnels) a whole series of anticipations formed as to what we were likely to find next. If this really was a mill, our reasoning went, there must have been a head-race channelling water in and directing the flow at the paddles of a waterwheel. This was a prediction that could be tested, and accordingly excavation strategy shifted away from exploring 'habitation' space in the chamber towards establishing whether or not such a head-race existed. The predicted channel was indeed found, entering the chamber (now the 'underhouse' of the mill) at a much higher level and thus providing the drop of water necessary to turn the wheel. A large slab of stone had been placed below the point at which water from the head-race was funnelled in, to take the force of water and prevent erosion of the floor. Taking levels along the course of the exposed head-race, underhouse and tail-race, and finding a consistent fall in height – together with other evidence – helped to confirm the identification of the structure as the earliest example of a horizontal watermill in Scotland (Batey and Morris 1992, Batey 1993).

This re-interpretation not only entailed a radical re-think of the agencies and materials associated with the use of the building. It also led to a shift from a site-specific to a more inter-site or landscape perspective, and a corresponding change in investigation strategy. A mill does not consist of a single building alone, but is part of a much wider complex of structures. Finding the mill building was just a beginning. A search began for the point at which the flow of water was diverted from a nearby stream. The former mill-pool was found on the hillside above the site: here the stream was dammed to build up and control the necessary head of water to flow through the head-race and turn the wheel. The tail-race returned the water into the same stream just below the mill, and from there into the sea. Thus results of the excavation could be connected with other parts of the landscape both upstream and downstream of the site.

Recording levels with a dumpy level or theodolite along the course of inferred water flows is a simple but important technique in the repertoire of archaeology of flow. The assumption is that an overall downhill gradient was required to keep water flowing and facilitate its control. Proving the existence of such a gradient along the supposed direction of flow does not give absolute confirmation that a particular interpretation is correct, but can often be taken as strong supporting evidence.

When traces of flow are found on a site, as in the Earl's Bu example, a subtle but important change occurs in our ways of seeing and thinking about it, and our ways of digging it too. There is partial re-orientation of perception of archaeological evidence. As embodied human beings we orientate our experience of the world in relation to the vertical stance of our bodies. We have a front and a back, a left and a right. We move forwards or backwards, or from side to side. We look ahead, or turn to look behind. If using a map or compass, the cardinal directions of north, south, east and west provide a more abstract frame of reference less dependent upon the position and movement of the body, making it easier for us to envisage our location relative to all other points, or even to envisage the landscape without us in it. All of that still pertains in archaeology of flow, but in this case there is a re-orientation of the spatial framework of archaeological analysis to accommodate a further axis, around which the evidence can then be organised and understood. This axis, crucially, is *direction of flow* (Tilley 1997: 113). Along that axis there are two basic directions, which override all others: *upstream* and *downstream*. This adds another dimension entirely to archaeological interpretation.

One way to experience this re-orientation of perception is to take a boat out on a river. As the boat pushes out from the bank it is very quickly caught by the current. The direction of flow moves the boat in one direction and presents resistance to movement the other way. To control the boat is to encounter and respond to and allow yourself to be partially oriented by flow. Although not possible on this particular site, taking a boat out is to archaeology of flow what walking the land is to landscape archaeology, especially as outlined by Tilley in his 'phenomenology of landscape' (Tilley 1997). Just as Tilley found his direction of walking to be oriented in part by the linear banks of the Dorset Cursus – to be discussed further in the next chapter – so a river quickly orientates the person in terms of its direction of flow. The same thing happens while excavating an archaeological site which has material traces of flow running through it. Emerging evidence itself brings about the restructuring of perception around the axis of upstream : downstream, and once that has happened it is difficult not to start interpreting the past use of the site in terms of the movement of flow. How that flow might have been manipulated by people in the past, or ways in which it might have impacted on aspects of their lives, are now critical questions to be investigated while excavating the site.

Most of the material structures of flow described in this book can be properly understood only in relation to direction and force of flow. On some sites there may be multiple flows which change course and shape through time, making the situation complex and dynamic. Even when flows are unidirectional, however, there are nearly always continuities and connections in both directions. What happens in one place can effect what occurs both upstream and downstream of that point (subject to limits imposed by gradient and other factors). This is another way in which archaeology of flow is different from conventional archaeology of solid materials; there are no such things as discrete and bounded 'sites'. Rivers are best thought of as assemblages of different kinds of materials, entities, forces, flows and energies – human and non-human – all inter-connected with each other (Bennett 2010: 23-8). Thinking in terms of flow makes us cease to think of sites as bounded entities, and leads us to seek out wider connections.

All this might seem obvious, but actually much archaeological interpretation is in what might be called solid material mode, and the dimension of flow is not always perceived to be important. The study of watermills, for example, often focuses on mill buildings and machinery rather than the more extensive systems of pools and sluices, head and tail-races, connecting the mill with nearby rivers and streams. In the case of aqueducts, Trevor Hodge notes that while attention has been focused on the architecture of aqueduct bridge arches, the flow of water along the aqueduct as a whole is often largely ignored. This led to the structures being considered more as isolated monuments rather than as functioning elements of a wider system, and to the recording of 'the dimensions of everything but only a sketchy idea of how the water flowed' (Hodge 1992: 4-5).

Flowing boundaries

A similar situation pertains to the study of medieval town boundaries. There have been important advances and insights in the study of earthen ramparts, stone walls, towers and gates (Schofield and Vince 2003; Creighton and Higham 2005). But the outer ditches from which material for the ramparts was generally dug out formed an important part of these systems of defence too. Sometimes these were dry ditches, but in many cases boundary ditches were filled with running water – indeed were often dug specifically with that purpose in mind. Numerous medieval towns, especially those located on rivers, were partially or completely enclosed by running water (see Lewin 2010: 279-80; Guillerme 1988). To get from inside to out, or from outside to in, travellers would have to cross lines of flow – which constituted material and symbolic boundaries of the town just as much as adjacent linear earthworks of earth or stone. The fact that such boundaries consisted of flowing water alongside solid material made them all the more vital and vibrant. They were living, working, flowing monuments functioning within landscapes of power.

When the Anglo-Saxon burh of Wallingford in Oxfordshire was built in the late ninth century, the settlement was laid out – its boundaries demarcated – not only by immense earthen ramparts but also by lines of flow. One side of the town was formed by the River Thames, centred on the river-crossing which gave the town its name. The other three sides of the rectangle were marked by a large ditch running on the outer side of the huge rampart, topped by a wooden stockade or stone wall. While great attention has been paid to the defensive and symbolic value of the ramparts – and the man-hours needed to create them or the number of men needed to defend them – little attention has been given to the value of the flowing water that ran through the adjacent ditches. Yet evidence is emerging of the great effort and the enormous amount of labour invested in order to ensure a working flow of water (see also Guillerme 1988 for a detailed account of the use of flowing water in northern towns of France during the medieval period).

The general slope of terrain down to the Thames meant that diverting part of the river to flow around the town was not an option. Instead a channel was dug to bring water from a network of streams several kilometres away. On reaching the town boundary ditches, the flow was divided by a double-sluice to run both ways, thus enclosing the town and outflowing into the Thames in two separate places (Grayson 2004; Edgeworth 2009).

This could only be accomplished through working with the local topography, but even then significant obstacles had to be overcome. In the north-west corner of the burh, for example, there was a rise in ground level of over 2 metres. The solution was to dig a section of the ditch much deeper (and the rampart correspondingly higher) than elsewhere on the circuit. The reason for creating such impressive earthworks, then, was not just defensive: considerations of flow also influenced rampart position and size.

Why should builders of burhs go to such lengths to ensure

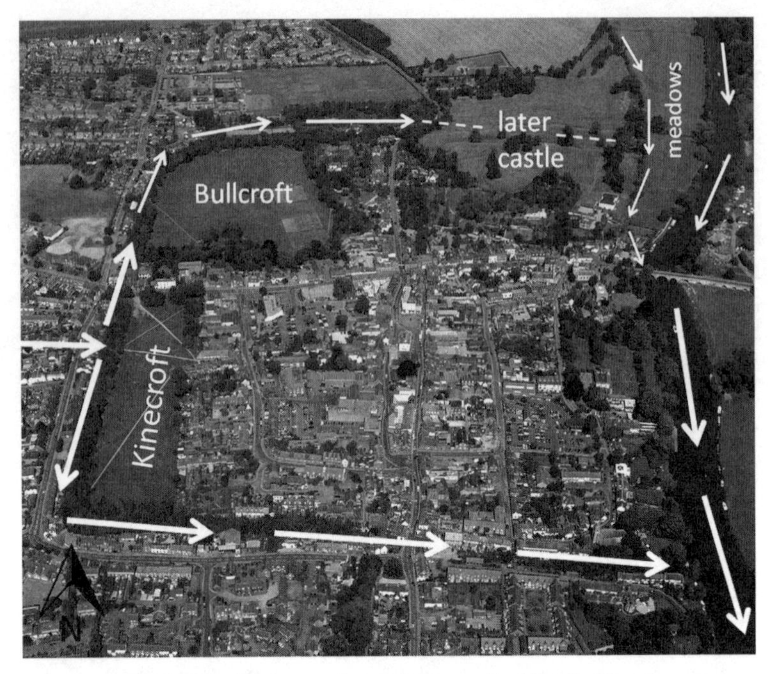

Fig. 4.1. Wallingford: a town bounded on all sides by flowing water. Aerial photo taken August 2001, © English Heritage NMR.

flowing water? An obvious reason is that stagnant water is of little use to anyone. It quickly clogs up with vegetation, leads to accumulation of black organic mud, starts to smell and becomes a breeding-ground for mosquitoes. It makes the earthwork difficult to maintain through periodic digging out of the ditch. Still water cannot be easily managed because it has no moving energy. Flowing water on the other hand can be readily controlled – by use of dams, sluices and diversionary channels. It can be put to multiple uses that stagnant water cannot, its energy harnessed for numerous industrial tasks. Crucially, flow can also be utilised to take away pollutants, residues, rubbish and sewage. A flowing stream is self-scouring, in the sense that the current will take away the finer particles of silt, leaving behind larger particles to create a well-sorted gravel bed, making the earthwork much easier to maintain. That is why such town boundary ditches were designed from the outset to be conduits rather than just containers of water.

The fact that the earthworks of Wallingford still survive to an impressive height and depth testifies to the continued usefulness of flowing water throughout much of the medieval period, when regular cleaning out of the ditch must have taken place. This is a multi-period monument, its flowing functionality periodically renewed right up to the post medieval period. But continuity of flow through time does not mean that things necessarily stayed the same. Existing flows were turned to new purposes, serving new masters in changing political contexts. Thus when Wallingford Castle was built shortly after the Norman conquest, the castle was constructed over the north-eastern stretch of the burh boundary. This was partly a matter of material symbolism – a powerful way of stamping a new authority over the old – but there was also a more practical and prosaic reason. Castle-builders needed a flow of water to fill their moats. An obvious recourse was to commandeer and make use of the existing flow provided by the Saxon boundary ditch. Excavations by Nicholas Brooks in the 1960s uncovered the former north gate of the town, built in the twelfth century over the partly silted-up town boundary ditch. The road coming into the town from the north crossed the ditch by means of a causeway, underneath which was a well-constructed stone culvert. The culvert brought the flow of water from the ditch into the extensive system of castle moats (Brooks 1966).

Another late Saxon town with a later Norman castle straddling the former boundary ditch is Bedford, and here too the commandeering of existing flows of water to fill defensive moats was a major factor in castle location. Like Wallingford, northern Bedford was situated on land with a gradual slope down to the river, making use of water redirected from nearby streams to flow around the town boundaries. But southern Bedford was a different matter. When the town was refortified in 915, a new burh was built on the south side of the river, turning Bedford into a 'double-burh' that enclosed and protected the river crossing. In this case, the gradient was flat enough to make use of redirected water from the river itself to flow in a semi-circle around the new fortified settlement – along the line of a rampart and ditch known as the King's Ditch. However, the gradient was too flat to ensure a strong enough flow to stop the water backing up and returning to the river the way it came. The solution was to build a weir on the river itself between inlet and outlet of the new earthwork, creating the necessary fall of water.

The line of a submerged stone barrage, stretching diagonally across the river, has long been known from documentary sources; it was recorded as having been demolished during navigation works in the late eighteenth century. Material traces of its former existence were last seen and mapped over a century ago as a broad band of weeds on the riverbed – roughly 3m wide and 40m long – making it of equivalent size to the Norman dam at Hemington already discussed. It was assumed to have vanished without trace. However, it recently reappeared in precisely the same location on Google Earth – the band of dense underwater vegetation easily visible from about a kilometre up on aerial photos taken when the water was low and clear.

The weir was more than just a feature in its own right. In creating the necessary fall to ensure flow through the King's Ditch, it formed an intrinsic part of the earthwork. Although not actually contiguous with it in terms of any solid connection, in fact it was connected with the monument through patterns of flow. It was probably part of the town defences in another sense too, for in blocking the river approaches to the town it would have served to restrict and control access by boat. There was little point in enclosing the town with earthen and water defences if river approaches were left undefended. The weir also provided the fall of water to drive a mill, and structured the whole river in terms of upper and lower levels. Later modifications to the river, such as the bypass channel (and the pound lock it probably contained), were adaptations to river

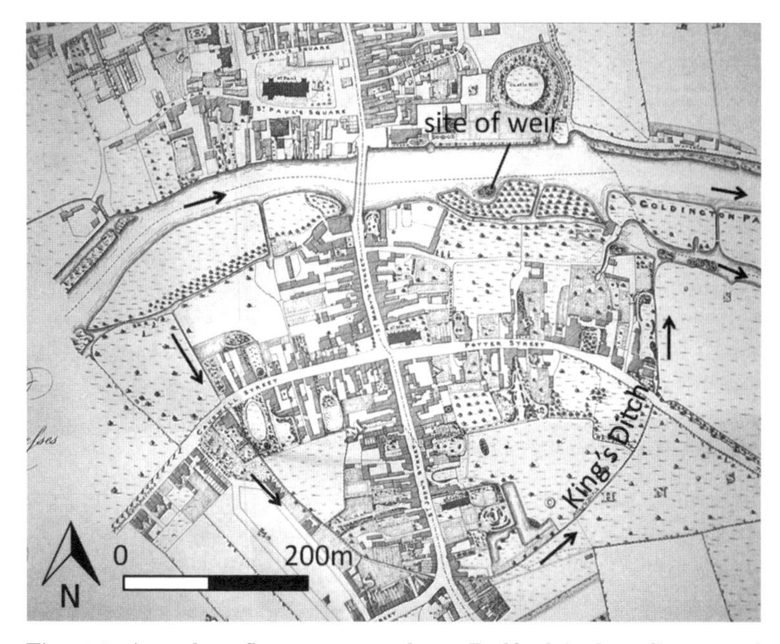

Fig. 4.2. An urban flowscape: southern Bedford in late Saxon and medieval times (Reynolds 1841 map used as base plan, with thanks to Bedfordshire and Luton Archives and Records Service).

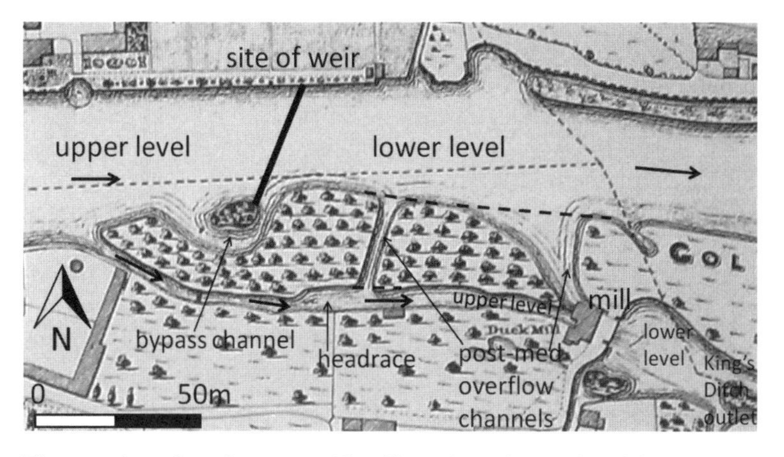

Fig. 4.3. An urban flowscape (detail): weir and associated features.

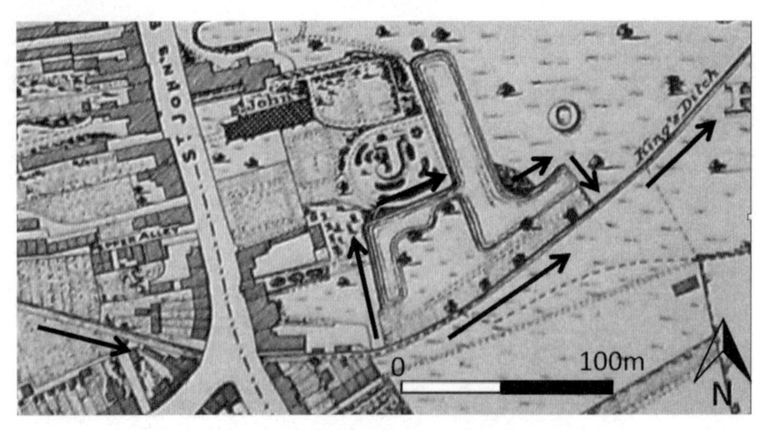

Fig. 4.4. An urban flowscape (detail): medieval fish-ponds.

conditions put in place by the earlier weir. Meanwhile the flow in the King's Ditch was also put to use in later periods, as illustrated by the complex of medieval fish-ponds, taking water from the ditch and returning it again.

Artificial distributary networks

In both Wallingford and Bedford, the channels that brought flow to the boundary ditches, the boundary ditches themselves, and later fish-ponds and castle moats that made use of these pre-existing flows, are all elements of what might provisionally be called artificial *distributary* networks.

It is well known that rivers consist of branching networks. Small rivulets on higher ground merge into larger gullies which in turn converge with successively larger watercourses that eventually flow as tributaries into 'trunk' rivers that flow into the sea. Such *tributary* branching networks are extensively described in books on hydrology and geomorphology.

But rivers can branch in the opposite direction too. 'Distributary' is a term normally applied in hydrology to any channel that branches off a main stream, especially in the context of deltas where a river meets the sea, or the alluvial fans that form where a stream exits from a canyon onto a plain. In this

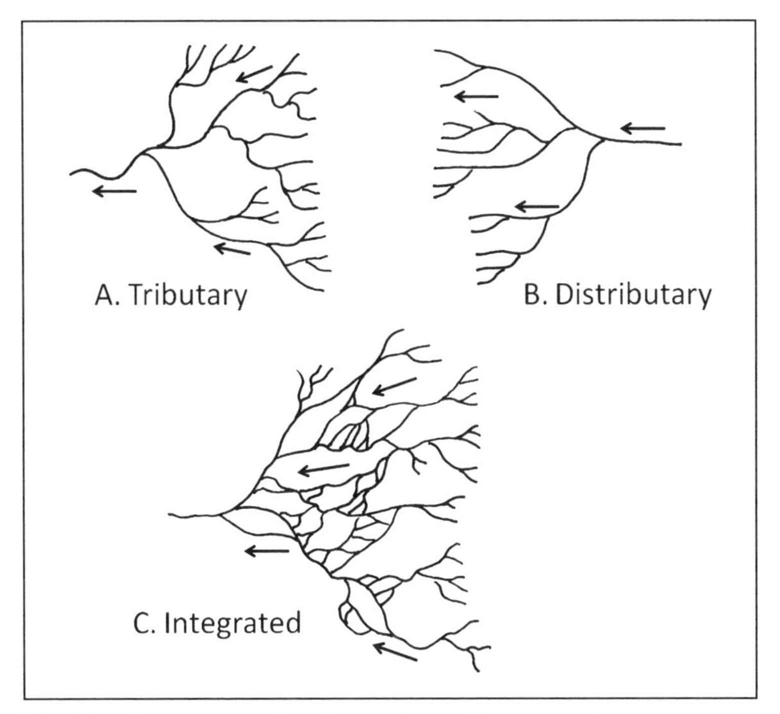

Fig. 4.5. Tributary and distributary branching networks.

chapter, however, the term is used in a slightly different way – to describe situations where water is artificially diverted off rivers or streams to flow into smaller watercourses, in order to serve particular functions connected with specific projects. Such channels may themselves branch into yet smaller watercourses to serve further purposes, with flow being distributed along ever smaller channels into the very fabric of urban and industrial life. These artificial *distributary* branching networks are often omitted from geomorphological accounts.

In reality, most rivers today are neither solely tributary (A) nor distributary (B) branching networks but integrated networks of water flow combining aspects of both (C), as shown in the schematic Fig. 4.5. For illustration of such an integrated system, see Muir's map of the Great Ouse, Nene, Welland, Cam and other rivers together with meander cut-offs and drainage channels in the southern fens of England (Muir and Muir 1986: 53).

Examples of artificial distributary networks can be drawn from both rural and urban landscapes. In early post-medieval irrigated water meadows in southern England, river water – and the fine riverine silt it carried – was distributed via a main channel into a series of subsidiary channels on artificial ridges, with water spilling from the carriers into a further network of furrows, and thence back to the river (Muir and Muir 1986: 115; Brown 2005). The rice-growing fields of China, Indonesia and many other parts of the world likewise depend upon complex systems of flowing water, fed by streams or irrigation channels, with water retained behind low mud banks on sloping hillsides and valley floors.

Some artificial distributary branching networks have already been encountered in previous chapters. The head-race of a mill, diverting part of the flow of water from a river, is an obvious and simple example. What is being distributed in this case, as well as the water itself, is the energy of the river, used to turn a mill-wheel to grind flour or perform some other task. The channel then becomes a tributary again when the water is returned into the river further downstream. Further channels may be taken off the mill race for other purposes, creating subsidiary distributary networks.

The part played by flowing water often gets missed out from archaeological accounts of industrial production. It is easy enough to see in the case of watermills – but what about, say, the making of pottery? This tends to be seen as a hot, dry process rather than a cold wet one. The role of fire is played up, at the expense of the role of water, which is played down. It is almost as if fire is regarded as the agency of cultural transformation, while water is seen as part of the untransformed natural world – a variation of the anthropological distinction between the raw and the cooked (Levi-Strauss 1970). The resulting pottery vessels and other artefacts are indisputably solid, and this makes it easy to forget the role of flowing liquid (and the mud carried with it) in their production. Perhaps for these reasons, most excavations of potteries tend to target kilns.

The excavation of a sixteenth- to seventeenth-century pottery production site I undertook for Birmingham Archaeology, on the edge of the historic core of Wednesbury in the heart of the Black Country, was targeted on a kiln to begin with. It was a surprise when a large linear water feature began to emerge from beneath dumps of pottery wasters. It turned out to be the town boundary ditch, alongside which the kiln had been deliberately located to make use of the supply of flowing water (directed from streams further uphill). Coming down a steep slope, the 4-5 metre-wide ditch was stepped into a series of different levels, widening out to form at least three settling tanks, presumably for collecting clay for processing. Smaller channels led from higher levels of the ditch into small rectangular pits next to the kiln (perhaps for mixing slip) then back towards the ditch to rejoin the flow further downstream. Many pottery production sites must have had similar facilities for flowing water, traces of which rarely materialise in excavation because they lie outside the immediate vicinity of kilns, on which investigation is targeted.

As in the example of the excavation of the Orkney mill, the practice of digging and interpreting such features becomes an exercise in archaeology of flow. Functions and meanings of the various water features only make sense when considered in terms of the gradient of the site and the direction of the flow of water in relation to other features such as kilns. The presence of material traces of flow animates the site and interpretation of it, adding a new and dynamic dimension to what would otherwise be merely a static configuration of features and artefacts.

An artificial distributary network providing crucial infrastructure for a late medieval town and still in use today can be seen in Gdansk, Poland. The Radunia Channel was constructed by Teutonic knights in the fourteenth century, modelling it partly on Roman aqueducts observed in Palestine, to bring water from the Radunia River to Gdańsk. Over 13 kilometres long, it hugged the contour line at the foot of hills, at a gentler gradient than that of the river itself, so that when it reached Gdańsk it was elevated several metres above the streets and buildings. The Grand Mill, located on the Radunia Channel in the centre of Gdańsk, made use of this contrived head of water to drive a total of eighteen vertical water wheels. It was one of the largest industrial installations in Europe, rivalling the mills of the Bazacle in Toulouse. Water was also released from the elevated channel through a series of sluices along the embankment into a network of wooden pipes, wells and cisterns, providing drinking water for the city – and eventually into sewage pipes which fed back into the river (Kowalik and Suligowski 2001).

Medieval hydraulic engineering facilitated the tremendous growth of the city, but also led to the devastating floods of July 2001. Large amounts of rainfall in the hills on the upward side of the Radunia Channel resulted in a rapid rise in water level and breaching of the embankment in five places on the downward side. The road running alongside the channel was turned temporarily into a major river, and the whole of the old historic core of the city, located in a depression so as to make use of water flow, was flooded (Majewski 2005). These events remind us that artificial distributary networks are still parts of river systems which have the capacity to act against human projects in unexpected ways, forcing adjustments and modifications of work carried out during previous interventions.

Another example of an artificial distributary branching network is provided by Abbé Vacandard's *Vie de St. Bernard* – a twelfth-century report on use of waterpower in a Cistercian monastery (quoted by Gimpel 1992: 3-6; Luckhurst 1964: 6). An artificial channel diverted water from a nearby river to the monastery. Here it was split into several smaller channels, one going all the way round the outside of the boundary wall, others irrigating gardens and orchards, but the main flow being channelled through a series of industrial installations. After turning the wheels of the watermill it went on to a brewing shed where it filled vats for making beer. It then powered the hammers of a fulling mill before going into the tannery. Then it 'dissolved into a host of streamlets' as it entered the inner court – the water being used for cooking, sieving, watering, washing and other tasks, before finally being returned to the river, taking with it all the sewage, industrial effluent and kitchen waste. As Gimpel points out, the same system with variations was used in most of the 742 Cistercian monasteries in Europe.

Bond (2001) describes the distribution of water through monasteries in much greater detail, right down to the network of pipes, cisterns, settling tanks, filters, sluices, drains and sewers. Such sophisticated systems of water control depended upon precision engineering and expert location of site from the outset. Many Cistercian monasteries were located in similar riverside locations for this reason, sometimes the site being extensively terraced before construction of buildings took place to facilitate the flow of water. It would be impossible or extremely difficult to construct flowing water systems like that on an ad hoc basis.

The same applies to the water that flowed through town boundary ditches at Wallingford and Bedford. From a hydraulic engineering point of view, initial placing of town boundaries must have been done with feasibility of water flow in mind. Ensuring water flow through the ditches was at least as important as strategic, military, political and other considerations in deciding the exact configuration and location of boundaries. Surveying of topography would have taken place beforehand and local undulations taken into account in earthwork design. This may partly explain why the ramparts in both places enclose an area somewhat larger than the settlement itself, with open spaces seemingly built into urban layout from the beginning. As André Guillerme says of towns like Caen and Beauvais in northern France, 'The medieval city was indelibly marked by water, subjected to its power, and scaled to its dimensions' (Guillerme 1988: 75).

Flowing water could be diverted from outer ditches or moats

to networks of intramural water channels, especially where – as often the case – there was a gentle gradient down to the river. Guillerme describes the technique of utilising, in former Roman cities, the grid of intramural Roman roads as water conduits. Bearing in mind that ground level had risen since the Roman period, the method was to dig a roughly 2 metre wide trench along the centre-line of the old road, down about 1 metre to the level of the clay roadbed foundation, which provided a watertight base. The flow of water along such channels could be used for a multitude of functions, including the powering of mills (Guillerme 1988: 57-60).

The many plans of intramural and extramural networks of flowing water provided in Guillerme's book, *The Age of Water* (1988) – with arrows indicating direction of flow and sections showing gradient of terrain – are recommended as graphic examples of the sheer complexity of artificial distributary networks in and around medieval towns, splitting up into myriad sub-networks of industrial supply, irrigation and drainage channels that eventually took water back to the river. Largely silted up and built over by the eighteenth century, such networks are generally forgotten about in discussions of water circulation in the modern urban city. The geographer's statement that 'the concept of "water circulation", with water following a given path into, through, and out of the city by the sewers remained foreign to western urban imaginations, spatial representations, and engineering systems until the 19th century' (Swyngedouw 2004: 30-1) is too one-dimensional a view. Although introduction of sewer systems was indeed a feature of the nineteenth century, there is a much greater temporal depth to the relationship between flowing water and the city than generally realised.

Artificial distributary networks extended through time as well as space – as successive generations made use of existing flows, adapting and modifying them to their own ends. Thus new channels were created to divert flow away from older channels, originally constructed for earlier water-related projects. At Earl's Bu there was evidence of later vertical mill wheels having been fixed to walls of surviving farm buildings, on a channel running off from the now-blocked earlier head-race – indicating that a succession of mills of different types utilised the same (modified) flow of water from late Norse right through to post-medieval times. This continuity of flow in midst of change is not untypical, which is why archaeologies of flow almost inevitably turn into multi-period investigations, exploring continuities and transformations through time from one period to another.

Early examples

Extensive artificial distributary networks are characteristic of the so-called 'hydraulic civilisations' of ancient Mesopotamia, Egypt, China, Sri Lanka, pre-Columbian Mexico and Peru, and so on). Historian Karl Wittfogel (1957) argued in his book *Oriental Despotism* that control of water on a large scale established the material conditions for the growth of early states. Drawing heavily from Marxist theory, he envisioned forced labour of masses of people by despotic rulers supported by complex bureaucracies. American anthropologist Julian Steward (1955) advanced a similar argument about complex systems of irrigation – using mainly water from rivers carried along artificial distributary networks of one kind or another – being a catalyst for development of state society.

In both Mesopotamia and Egypt, forms of basin irrigation were practiced, in which a complex of earthen banks formed basins of various sizes to retain floodwater, which was then released through sluices into further basins or into canals and irrigation ditches. The system was adapted to the very different conditions in the two regions. The River Nile usually flooded in summer, with the inundation of the floodplain gradually falling back in time to plant crops in the rich newly-laid alluvial deposits in Spring (Brown 1997: 5-13). In Mesopotamia, the Euphrates and Tigris were more susceptible to flash floods at the wrong time of year, leading to the development of much more complex networks of canals and irrigation ditches to retain and distribute water. Showing that rivers and other watercourses need not be the dark matter of landscape archaeology, Wilkinson (2003) gives an excellent account of traces of flows in archaeological landscapes of the Near East visible today.

Steven Mithen provides a summary of the works of other hydraulic civilisations – at Khmer Angkor (Cambodia), the Aztec city of Tenochtitlan (Mexico City), Constantinople (Istanbul) in the Byzantine Period, and Petra in Jordan (Mithen 2010: 5251-5). As Mithen points out, one could go on to talk about the hydraulic achievements of Mycenaean Greece, the Harappan civilisation of the Indus valley in India and Pakistan, the Mayan and Inca empires – and still not have covered anything like the vast extent of the subject. It is only possible to skim the surface here.

Worthy of note because of their widespread historical use in hot arid areas are the underground canals known as *qanats* (Iran) or *foggera* (North Africa). Originating in pre-Islamic Persia at least 3,000 years ago, their use spread throughout Arabia and the Middle East, Pakistan, India Afghanistan, Spain and even (through colonial introduction) South America. Constructed as a series of well-like shafts, connected by gently-sloping tunnels, qanats transport flowing water from upland aquifers over long distances, using only gravity and minimising evaporation loss (English 1968; Beekman et al. 1999; Wilson 2006).

Qanats are really underground aqueducts. We tend to associate the word aqueduct with the great arched bridges that took conduits of water across a valley, but actually aqueducts can be in the form of tunnels, bridges, arcades, canals, dykes, inverted siphons, and so on. Like qanats, aqueducts worked by gravity. Bringing water into cities, towns and industrial sites, Roman aqueducts provided a continuous supply of water for public baths, drinking, public fountains, mills and other industry, and ultimately for carrying away waste, usually into the river. When approaching lowland cities on great aqueduct bridges or arcades they also served as giant status symbols. Even cities like Lugdunum (Lyon) in Gaul, located on the confluence of two major rivers, had four aqueducts heading towards it. One of these, the aqueduct of the Gier, was 85 kilometres long. For much of this distance the concrete culvert ran close to the surface, but there were also tunnels, arcades, bridges and inverted siphons – the full range of building techniques being employed to maintain the overall sloping gradient along its entire length (for a full account of principles of flow engineering entailed in the building of Roman aqueducts, see Hodge 1992).

Up to now, I have used the term *artificial* distributary networks to make the distinction with the *natural* distributary networks that are dealt with by hydrology and geomorphology (such as those that form on deltas). In reality, the distinction is not nearly so clear-cut. In the Fertile Crescent region during the fourth and third millennia BC, for example, the Tigris and Euphrates Rivers often breached levees to form 'crevasse splays', resulting in a fan of coarse sediment being deposited on the floodplain and the splitting of the water that had broken through into multiple channels – a classic distributary pattern and might seem to be entirely natural. However, in some cases the levees were deliberately breached for irrigation and other purposes (Haevaert and Baeteman 2008; Wilkinson 2003: 88-9). Floods might then have widened further those humanly wrought breaches, sometimes bringing about an avulsion or sudden change in river course. Another scenario was when natural breaches were artificially widened to deliberately encourage flowing water onto the floodplain. Were these natural or cultural events? Returning to the argument of previous chapters, the answer would have to be neither one nor the other, but rather entanglements of both.

Many so-called canals or artificial waterways were actually modifications of natural channels to start with, making use of former river courses or straightening existing ones. Even if entirely man-made, artificial channels were subject to the same forces of erosion and deposition as natural watercourses, often becoming 'naturalised' over the course of time. Thus a once broad canal may largely silt up, the remaining flow of water taking the form of a meandering stream (though edges of its meander belt may still be defined by the original banks of the canal, giving a clue to its formerly artificial status). As we have already seen, flow does not really fit into opposed categories of artificial and natural. Tempting as it is to classify as one or the other, within frameworks of thought built upon such dualisms, nature and culture converge and intermingle in all the examples cited in this book.

Even the most natural-looking distributary patterns which form on deltas are often actually entanglements of natural and cultural forces. Sometimes the very locations of deltas have been influenced by human action further upriver, either through artificially induced change of course of the river, or the artificial prevention of change of course. If the Mississippi is an example of the latter, the lower reaches of the Tone River in Japan would be an example of the former. In the sixteenth century most of its flow was diverted from its mouth in Edo Bay to flow into the Pacific 100 kilometres to the east, facilitating the drainage of swamps and allowing Tokyo to develop into the urban centre it is today (Uzuka and Tomita 1993). In any case, as we have seen, the amount of sediment deposited in deltas is itself to some extent an artefact of different kinds of human intervention (notably the building of dams which act as sediment traps, heightening of levees constraining deposition on floodplains, and deforestation which increases erosion and therefore the load of material carried by the river to be deposited in delta regions).

In the light of the above discussion, then, the distinction between 'artificial' and 'natural' is clearly just as problematic in describing distributary branching networks as tributary branching networks, which in any case are integrated together into single systems of water flow. A holistic view of rivers would include the whole spectrum of both tributary and distributary patterns, all understood as to some extent entanglements of natural and cultural forces.

A common tendency in archaeology is to look at 'artificial' distributary networks as discrete systems, removed from the broader river systems and water cycles of which they are a part. That leads to the 'water management' model, with emphasis on cultural control, and a view of water as a passive resource entirely subject to human will. As Mithen puts it, 'Water has been domesticated. By this I mean that its natural properties have been constrained and manipulated to cater for human need' (Mithen 2010: 5250). While there is much truth in this statement, there is something missing too, not least the sense of water as a vibrant material in its own right and vital force in landscape change. If water management systems were studied only in terms of domestication we would inevitably lose sight of the river response to human action, the sometimes wild and unpredictable aspect of flowing water, and the crucial dynamic that unfolds between humans and flowing water that is a principal theme of this volume. Archaeological evidence of the age-old 'wrestle' of natural and cultural forces would be overlooked or misunderstood. It is almost as though we can only bring rivers into the domain of cultural analysis by denying their non-human aspects and by rendering their flow passive and inert. A crucial step, then, is to re-establish material connections between so-called water management systems and the partly wild rivers or other water sources from which their flow has been taken, and to which it is returned.

Summary

This chapter has looked at examples of flow having been incorporated into the material design of sites and layouts, and some of the practical implications this has for archaeological investigation, as well as for our understanding of rivers. The obvious fact that water flows downhill, with certain exceptions, gives archaeologists a basic principle on which to base inferences and interpretations. The concept of artificial distributary branching networks was introduced, and it was pointed out that these are just as much part of river systems and water cycles as the more familiar tributary branching patterns. Rivers need to be remapped and rethought taking artificial distributary networks into account (though my own use of the term 'artificial' to describe them has been called into question). In exploring cultural dimensions of rivers and reintegrating the dark matter of landscapes back into archaeological study, archaeology of flow is more than just another subfield separate from the main discipline. It emerges from, and can be integrated back into, the normative archaeology of solid materials and landforms.

5

Land flows

Introduction

Starting with rivers, this book has gone on to consider other material forms and layouts that incorporate or make use of water flow. In this chapter it will be asked – can perspectives and techniques of archaeology of flow be applied to landscapes away from rivers, and to the archaeology of sites that seemingly have little to do with water as such? What other kinds of flow run through landscapes? How do these intersect with rivers and associated flows of water examined so far?

The answer to the first question is yes. Patterns of flow occur not just in water, but also in other parts of the natural, technological and cultural worlds. Flow manifests itself in moving air, shifting sand, ever changing cloud formations, stampeding animal herds, swarming bees, shoaling fish, and flocking birds (Ball 2009a). It can also be found in music, spoken and written language, electric circuits, computer networks and information superhighways. When we speak of 'streams of information' or 'political currents', we are partly using metaphors: but we are also expressing something real about how things move and change. Information really does flow. Systems of ideas really do run in currents. Traffic really does stream. People moving through crowded station concourses or university quads really do move in flows. The frenetic buying and selling of money in a stock exchange really can on occasions exhibit turbulence. As anthropologist Anna Tsing puts it:

> Imagine an internet system, linking up computer users. Or a rush of immigrants across national borders. Or capi-

> tal investments shuttled to offshore locations. These worldmaking 'flows'... are not just interconnections but also the recarving of channels and the remapping of the possibilities of geography (Tsing 2000: 327).

There have always been flows and currents of one kind or another in human life, and physical traces of these are manifested in artefact assemblages, landscapes and townscapes at different scales of analysis. In referring to global flows on a large spatial scale, Tsing is simultaneously looking at relatively short time-scales. But flows of materials can occur on much smaller spatial scales and much longer time periods too. In his examination of the archaeology of personhood, Chris Fowler examines the 'flows of substances' in gift exchange that can cross the fluid boundaries of persons, and the flows of ancestral substances that may be understood to pass from body to body across generations in mortuary practices (Fowler 2004: 20, 67). Material property can flow as heirlooms through many generations along paths of descent, and even apparently solid material like stone can flow from one building to another over periods of hundreds of years through multiple phases of use, demolition and re-use.

Paths of ancient flows that pulsed through people's lives hundreds or thousands of years ago can be found in the archaeological record. Indeed, it is difficult to make much sense of most archaeological evidence without understanding it at least partly in terms of fluid movements, eddies, currents and flows.

Roads as channels and conduits

Roads are obvious examples of non-water landscape features that are very much to do with flow. In many respects roads behave like rivers, sometimes taking a meandering form that is straightened out from time to time. It is well known that motor traffic on roads today can be analysed and made sense of (and thus controlled) by applying models of flow, which assume that the traffic behaves like a continuous fluid. But such models are not just relevant to the movements of cars and trucks: flow has been part and parcel of travelling since long before the motor engine was invented.

Consider for example the medieval hollow-ways that streak the hillsides and plains of many parts of Europe (Hindle 2008: 8-17; Muir 2000: 95). Most of these roads have never been constructed as such: rather they have been eroded by flow. In this case it was the flow of people, their animals and carts – not water – that was the primary force in their creation. Yet so similar were they to rivers and streams in their basic form that they often became water conduits too. Water erosion speeded up the hollowing-out process, and made some lanes muddy and impassable. When this happened alternative routes were taken by the people travelling along them, leading in some places to formation of river-like patterns of braided channels, branching and converging. Such features could be said to be half-rivers, half-roads – an intermingling of natural and cultural forces. But actually they are outcomes of much more complex assemblages of material flows (movements of people, animals, soil, dung, wheeled vehicles, goods, rain, wind, channelled water, cultural traditions, and so on), all mixed up together within dynamic processes of landscape change.

Hollow-ways are a type of archaeological feature found throughout the world. Many are still in the process of formation. Wilkinson (2007) compares present-day patterns of cattle trails around villages in Mali and Ivory Coast (West Africa) with tracks radiating out from Near Eastern tell sites of the Early Bronze Age. Not all should be seen as being wholly cultural phenomena; they have been partially shaped by water as well as by human and animal flows, and are thus enmeshed to some extent with local hydrology. Wilkinson gives instances of how hollow ways intercepted flowpaths of water across the landscape. In one example he describes how water was channelled by hollow-ways away from the central tell settlement site by radial paths down into meandering wadis or streams in the surrounding landscape. In another example, hollow-ways bring water and alluvium in towards the settlement from the surrounding area, creating a curious 'halo' effect around it on aerial photos (Wilkinson 2003: 112; 2007).

All this goes to show the extent to which flowing water is inextricably intermeshed with landscape features that might at first sight appear to have nothing to do with water. But let us leave this aside, and stick to the idea of roads as channels for human flow. Conceiving of roads in this way involves a subtle shift in the way we look at towns and villages. Instead of focusing on solid structures made of brick, stone or mud – the built fabric of a settlement – our attention shifts onto *the spaces in between*, through and along which movement occurs. It also changes our view of structures along the course of roads. Lines of buildings on either side, whatever other purposes they may serve, also function in part as banks or sides, directing traffic flow along a fixed and stabilised course. When a broad town street comes to be infilled with buildings, either through organic growth or as part of a deliberate planning exercise (a common occurrence in medieval and post-medieval towns in Britain), these present an obstacle to flow as surely as weirs or dams in a river. The flow now has to find its way through a network of smaller streets between and around the infilling buildings. The speed of flow may slow down, but the density of flow will increase, as more people try to pass through constricted spaces: flows coming in from different directions may create congestion (as in a crowded market-place) and even, on occasions, turbulence (as in violent political demonstrations). Gateways obstruct and facilitate flows as surely as sluice-gates do. Alleys, arches, tunnels, stairways, corridors and passages are conduits of flow as surely as streams or rivers are.

Of course the flows referred to here are comprised mainly of conscious, thinking, reflecting individuals – taking with them their vehicles, animals, goods, artefacts, etc. – embarked upon specific errands for particular reasons. Unlike running water, people are not restricted by gravity to a single direction of flow; they can move uphill as well as down and in more than one direction at the same time: indeed, look at any crowded stairway and you will see that they often organise themselves into two streams of traffic heading different ways. This may partly come about just by virtue of each person trying not to bump into anyone else. Yet clearly there are other more subtle influences at work too. Written notice-boards, direction-markers with arrows or other signs on, kerbs, lines marked out on the road, maps, spoken language and gestures from other people – as well as physical channels and barriers – may all have significant effects on movement of persons. Intangible cultural rules and etiquette might also restrict or direct movement along particular lines – as, for example, in a slow-moving flow like a queue (and here it is worth considering contemporary material layouts that are connected with queues – from turnstiles at football stadiums to the zig-zag lanes made from moveable barriers linked by extendable plastic tape, used to form and process flows at airports, banks and post offices).

On the other hand, people are making creative decisions from moment-to-moment about where they want to go, sometimes ignoring rules and resisting popular currents to move against dominant flows. These are the 'wandering lines' or 'efficacious meanderings' of de Certeau (1984: xviii) that undercut the strategic designs of the planners of social space (Ingold 2007: 103). Paul Graves-Brown (2007) describes how contemporary flows are channelled along particular routes through suburban shopping centres, leaving 'non-spaces' in the designed landscape where people are not supposed to walk. But he also points to the informal paths through hedges and flowerbeds transgressing such negatively constituted areas despite the attempts of those in authority to block entry.

> Normally speaking we tacitly accept the many boundaries and non-spaces that are created in the urban landscape, but when the sanction of these boundaries becomes inconvenient we overcome our tendency to conform (Graves-Brown 2007: 80).

Motor traffic is a more regulated kind of movement, yet one in which road signs, flyovers and underpasses, traffic lights, decisions, gears, oil, plans of journeys, tyres, embodied skills, tarmac surfaces, anticipations, changes of mind, and so on, are all mixed up together in a dynamic entanglement of materials, surfaces and ideas. It is the sum total of all these things and processes that produces the collective flows and counter-flows channelled through layouts of towns and villages. By analysing material layouts, something of former patterns of flow – perhaps even the rationales and ideas of the people who planned and controlled them – can be deduced. Something of the experience of people moving through a townscape can be reconstructed.

Flows of people through landscapes can be quite extended in time and space. Consider for example the medieval pilgrimage routes across Europe to Rome and other religious centres. Hundreds of miles long, some of these routes were trodden by pilgrims for well over a thousand years, and are still in use today. The stream of pilgrims not only linked together existing towns and villages; it also stimulated development along the route – providing the rationale for increased trade and commerce, further urban foundation and growth, building of churches and shrines, setting up of hostelries to serve the needs of travellers, and so on. The existence of flows over such broad sweeps of time and space presents opportunities for archaeologists to study landscape as something other than static, solid landform.

In an archaeological study of the Camino de Santiago de Compostela in northern Spain, Julie Candy explores experiences and life-worlds of medieval pilgrims who travelled by foot and pack-animal along the track towards Santiago (Candy 2009; 2005). She argues that the Camino is 'so much more than a line on a map which links dots signifying individual "sites" or landmarks along the way'. Instead the track has a clear directionality to it, which both results from and generates flow. It is envisaged as 'a sequence of places, unfolding through time and space' and 'a succession of experiences: of sights, smells, remembrances and associations that come to mind via the walking body within a dynamic, resonant landscape' (Candy 2005: 4).

This is getting close to the kind of moving perspective that Mark Twain, in piloting a steamboat, took up on the unfolding and shifting shape of the Mississippi. Landscapes that might otherwise be visualised from stationary points of view – as in maps and landscape paintings, for example – look completely different when reconfigured by the flow of the observer's own embodied and situated movement through them. A pilgrim on the Camino passing through the town of Puenta La Reina, at the end of the main street,

> would emerge from the dark, narrow, enclosed space up onto the high six-arched bridge ... The sense of space and light on the bridge, in contrast to the gloom, noise, confinement and tunnel effect of the town, marked another transition as the pilgrim continued on his or her way (Candy 2005: 6).

Here it is not so much monuments and sites that are taken to be of the greatest importance, but rather the journey *through* and *past* such places – the transitions between them in the context of a larger movement or flow. In this respect, towns and villages and buildings are understood, not just as places in their own right, but also as landmarks, ports of call, resting-places, half-way houses, refugios, or 'stops' *en route*.

An important part of Candy's method was to walk the Camino herself, in order to get that sense and perspective of an embodied, situated perceiver moving through the landscape or townscape. The approach is partly inspired by the phenomenological method of Christopher Tilley in walking the Dorset Cursus (Tilley 1997). But there are important differences. In the case of the Camino there has been a more or less unbroken if constantly changing tradition of pilgrimage along the route – a continuity of flow – since the medieval period. Tilley, on the other hand, had the more difficult task of envisaging flows that ceased in prehistoric times. The Dorset Cursus, moreover, has largely vanished from the landscape as seen from an embodied perceiver on the ground. It is best seen on maps and aerial photographs, which Tilley referred to as he walked. Like roads and other channels of movement, the linear form of the Dorset Cursus, with high banks on either side, imparted a direction upon movements and perceptions of embodied individuals traversing the landscape. It was literally a 'conduit of movement' (Tilley 1997: 199).

Tilley's version of phenomenology, while influential, has drawn considerable criticism (Fleming 1999; Brück 2005). Perhaps the most controversial part of his argument is the assumption of a partial correspondence between the experience of a person walking the route today and someone walking it thousands of years ago. This implies, according to Ian Hodder and Scott Hutson (2003: 119), the positing of a 'universal body' that denies the cultural specificity of human experience. Yet this is to slightly miss the point. It is true that there is no such thing as a universal body, for human bodies are socially and politically as well as naturally constituted, and are always historically situated. What Tilley succeeded in doing was to shift the terms of reference for theoretical issues like these, overturning the static viewpoint of much archaeological theory and bringing the flow of movement back into the archaeological perspective itself – *interpreting the landscape from a flowing and moving point of view.*

Many theoretical issues are transformed when they are put in the context of movement, and the question of the universal body is no exception. Even allowing for variously shaped bodies, cultural variations in posture and gait, different historical circumstances, and accepting that walking is a learned, social activity as well as a partly genetically ordained predisposition, the bipedal form of locomotion presents certain possibilities and constraints which to some extent can be regarded as commonalities of human experience. The pull of gravity upon the body when walking uphill, the shift in stance that is necessary when starting to walk downhill, the way it is necessary to lean slightly back rather than forwards so as not to fall headlong down the slope, the changing position of hills on the horizon in relation to the eyes if the head is turned around to view the landscape from an upright walking stance, and so on – these are all examples of flowing movements that would have been part of walking the Cursus in Neolithic and modern times alike, despite the great changes in the landscape and the very different cultural contexts in which it is perceived and interpreted. Tilley was right to note how material structures in the landscape can channel movement and thereby structure the flow of perception of people (including the archaeologist) moving through.

In her article 'Landscapes on-the-move', Barbara Bender argues that the ways in which people on-the-move engage with landscapes are neglected in landscape studies (see Urry 2007 for the wider point that issues of mobility have been left out of social scientific debate in general). Referring to nomadic tribes in Mongolia and contemporary Roma in Europe, Bender strongly challenges the view that such people are 'dis-located' or 'out-of-place'. On the contrary, they shift the central axes of their worlds every time they move tents or caravans. Far from being unattached, their attachment is to a moving, shifting landscape rather than a static one (Bender 2001). This is not to deny, of course, the very real sense of dislocation that refugees and other displaced persons may feel.

Pastoral nomads may leave little or no trace in the archaeological record, though that is not always the case. Certainly their footprint is lighter than that of sedentary groups, and Wilkinson gives the example of Omani sheep-herders who simply live beneath a convenient tree (Wilkinson 2003: 173). But 'nomad architecture' can be distinctive, with camps containing numerous specialised areas and fixtures in characteristic configurations. In theory it should be possible to reconstruct paths of nomadic groups from the distribution of such material traces across landscapes – by 'following their tracks', as it were.

Even landscapes of those who stay put – such as town-dwellers and sedentary farmers – are also on the move. Bender alludes to contemporary 'forces that pass powerfully through – television, radio, tourists, commodities, armies' (Clifford 1992: 103), not to mention digital, mobile phone and other emerging technologies. But as we have seen, powerful flows and currents passed through medieval and prehistoric worlds too. Landscapes change, and not just as the result of human action upon it. Changing, flowing landscapes are prime movers and shakers of human life and always have been. By switching our attention from solid and static landforms onto the 'fluxes and flows of materials', as Ingold (2008: 3) entreats us to, we begin to glimpse the 'creative entanglements' of people and landscapes.

Funnels

Landscape flows are bound up with movements of animals as well as people – often both together. Animal husbandry is of central importance to pastoral and nomadic societies, and until industrialisation the same was true of countries like Britain and the United States. Herds of animals were driven short distances between farms and markets on a network of local drove roads, and taken on much longer journeys on larger transhumance routes that crossed vast tracts of countryside. Although patterns of flow varied from place to place, animals were typically taken to upland pastures in summer and brought back down to lowland valleys in winter, often over long distances.

Cattle, sheep and horses are all herd animals by instinct that, when herded together, move in a flowing manner. Somewhat different to the interaction of particles of running water, animals in a herd each respond very quickly to movements of neighbouring animals, giving rise to flow-like patterns. Look at a herd of cattle being driven quickly through a narrow space and you will observe currents moving at different speeds within the overall flux, eddies that form and disappear, currents that spiral or eddy back around the side to rejoin the general flow. Bottlenecks will form as animals try to squeeze through constricted spaces, creating turbulence and waves – as the forelegs of some animals climb up onto the haunches of others – slowing down one current and speeding up others on either side.

Herders (along with the horses they ride, or the sheepdogs to whom tasks can be reliably delegated) are skilled at manipulating flow in relation to particular sets of material layouts. Many kinds of archaeological structure were designed to enable, channel, contain or obstruct currents and flows of animals. It becomes easier to recognise and interpret such structures when it is realised that control of flow is a primary function. Thus Francis Pryor describes Bronze Age field systems in Eastern England, the central arteries of which were drove-roads feeding into and out of stockyards, fields, or other kinds of enclosures. Funnels formed by curving ditches, hedges and banks on either side were constructed and used by herders to direct animals through gates into corners of fields, and thence through a series of narrow linear pens and paddocks where animal sorting and exchange could take place. These were landscapes of flow, central to the economy and the community. As a sheep farmer as well as a well-known archaeologist, Pryor interprets them not as static structures or layouts, but as embedded material components of dynamic farming and social practices, with animals being moved around the environment in controlled flows and streams (Pryor 1996).

The droveways that Pryor refers to above are little more than 10 to 20 metres wide. In central Italy during the medieval period, however, shepherds and their huge flocks of sheep traversed a network of *tratturi* or drove-roads which had a standard width of 111 metres – these often being situated within even broader linear corridors of land up to 1 kilometre wide. Smaller droveways called *tratturelli*, up to 40 metres broad, linked the main routes. These in turn are linked by still smaller droveways. Every year, great rivers of sheep were taken along *tratturi* to upland pastures in Abruzzo province in summer and brought back to the plains of Puglia in the autumn – a distance each way of several hundred kilometres – leaving behind a veritable archaeology of flow. *Tratturi* boundaries were rigorously measured and marked out to exact specifications (presumably with ditch and bank, though excavations are needed to verify this), testifying to the fact that these flows were tightly managed and controlled.

Such broad linear spaces for flow were more than just tracks and routeways; they were linear pastures, providing grazing as well as passage for the animals. Immense herds traversing the *tratturi* were watered by linking routes in with springs and ponds at key points along the way, with streams specially diverted to create an interweaving pattern of land and water flows. Heavily regulated and under royal control in the Middle Ages (providing the basis for a lucrative wool trade), the transhumance system was still partially in use even in the early twentieth century. Its origins are thought to go back to Roman and prehistoric times (Christie 2008).

Similar systems of broad medieval droveways are known in Spain, where they were called *cañadas* and *cordeles* (Gómez-Pantoja 2004: 94). Britain too had its extensive droving routes and traditions right up to the late post-medieval period. Nothing of the scale and organisation of the *tratturi* and *cañadas* is known, but the former existence of such highly regulated systems is suspected. Landscape configurations of a similar scale are the funnel-shaped greens, commons and linear meadows studied by Susan Oosthuizen (1993, 2002). Still to be found fossilised within patterns of field boundaries today, these were integral parts of early medieval landscapes in Eastern England, though possibly originating much earlier as part of Roman or Iron Age transhumance practices. Whatever their date of origin, the characteristic funnel-shape can be taken to be indicative of flow through the landscape – the flow of animals under human control. Similar funnel-shaped configurations can be found in towns, especially leading in and out of marketplaces or river-crossings, and here it is clearly human as well as animal movement that was, and in many cases still is, being funnelled.

Less formalised trails formed by the movement of animals can be found in many parts of the world. Herds of up to 10.000 Texas Longhorns at a time were driven north from the Red River along the Chisholm Trail to supply beef to Kansas railheads – and from there by train to army posts, Indian reservations and industrial centres throughout the North after the American Civil War (see the 1948 film *Red River* with John Wayne). The erosion caused by such vast numbers of animals must have seemed to rip the landscape apart. Now mostly erased by later ploughing, the many branches of the trail, and others like it, can still be traced in the landscape, especially in Oklahoma. Here most of the cattle followed the same route north from the Red River, with deeply rutted broad trackways funnelling in to ford streams and rivers at established crossing-points.

A focus on flow brings to light new classes of monuments, or old ones simply not considered significant before. Pounds, shepherds' huts, and sheepfolds are cases in point. Until the landscape artist Andy Goldsworthy started his extraordinary sculptural work on sheepfolds (as illustrated in his 2007 book *Enclosure*) these were the least fashionable of archaeological structures. Yet Oscar Aldred's account of *réttir* of Iceland makes clear the potential of studying human activity and social organisation (not to mention entanglements of humans, animals and environment) through archaeological examination of animal enclosures. *Réttir* are large sheepfolds constructed from stone or turf, but – as Aldred explains – so much more than simply for gathering sheep. In Iceland the term *rétt* means the social occasion of herding animals (a seasonal festivity and assembly of people as well as a farming activity) in addition to the actual material enclosure where the gathering takes place. In other words, these are monuments for the gathering of whole communities. In autumn sheep and horses are herded down from the upland summer pastures and brought into the central enclosure of the *rétt* to be sorted into the relevant stall – each of which belongs to a particular farm. Thus the layout of the *rétt* is both material reflection and reproducer of social organisation (Aldred 2009a, 2009b; Aldred and Madson 2007).

In representing the movement of sheep through the *rétt*, Aldred uses arrows to illustrate direction of flow. In describing the monuments, he draws attention to the funnel-like spaces from outer to inner enclosures, simultaneously enabling and constraining flows of animals. As he puts it:

> Sheep folds are markers of movement in the sense they are places that there is a gravity of movement towards … But they are also makers of movement, through flows that occur through the monuments (Aldred 2009).

Flow always has this double-aspect to it. Its form, density and speed is shaped in part by material channels through which it is directed, but at the same time it in turn shapes those very channels. Rigidly managed and controlled, flow can serve to reproduce social values and practices. Left to itself, however, flow is potentially destructive and will carve out new channels in unpredictable ways (as in a flood, or an animal stampede). For this reason, human and animal flow may be subject to strict regulation by order of law and payment of tolls, as well as by reinforcement of banks, hedges, ditches, fences and other material boundaries that channel and funnel movement.

Many examples of traces of controlled animal flow can be found in prehistoric landscapes too, and in arid as well as temperate parts of the world. Large oval stone enclosures called 'kites' (because of their resemblance to kites when seen from the air by pilots) are distinctive features of the desert in Syria, Jordan, Israel and Saudi Arabia. These have pairs of long antennae-like walls (having the appearance of kite-strings) stretching out across the terrain, often for hundreds of meters – interpreted by Wilkinson (2003: 175-6) as funnels through which great flows of gazelles and other wild migrating animals such as ibex, oryx or onager were herded towards a central kill-zone. Sometimes kites are linked together in complex chains across broad paths of annual migration, often at places where obstacles such as cliff-edges created natural funnels or bottlenecks in the flow of animals, increasing their number and density. In this sense the structures functioned rather like dams built across watercourses, diverting or funnelling the flow of animals into side-channels. In many cases the walls were not high enough to stop animals leaping over, yet succeeded in channelling them because the funnel was constructed so as to always appear as an opening in the walls, at any given point of the animals' passage through, with the ultimate destination of the killing zone hidden from sight behind ridges until the last moment (Holzer et al. 2010).

The construction and use of such structures appears to extend through thousands of years from recent times back to the pre-Neolithic (Helms and Betts 1987; Fowden 1999). They are evidence not only of animal migration routes, but also of movement of groups of nomadic or semi-nomadic hunters who followed or intercepted the annual movement of animals – that is, of human as well as animal flows (or rather the entanglement of both). As with rivers, these material traces of flow testify to a complex intermeshing of natural and cultural forces over long periods of time.

Crossings and gatherings

Crossings of land-flow paths over water-flows (bridges), or water-flows over land-flow paths (fords/wades) are key locations in landscapes of flow.

The philosopher Martin Heidegger recognised that bridges do more than simply cross a river or make a connection between its two banks. 'The bridge,' he pointed out, 'gathers the earth as landscape around the stream' (Heidegger 1971: 152), reconfiguring the landscape in a new way. As Julian Thomas has shown, Heidegger has much to contribute to archaeological theory (Thomas 1996). But Heidegger is clearly wrong to insist that the gathering location never precedes the bridge. When he says, in that famous phrase, that 'the bridge does not first come to a location and stand in it; rather, a location comes into existence only by virtue of the bridge' (Heidegger 1971: 153), he overlooks the fact that in many cases bridges were built on the location of pre-existing fords.

A ford gathers the landscape around the stream in the same way as a bridge, though in a more fluid, less rigid and less visually imposing manner – requiring little or no investment in infrastructure. Like a bridge, it pulls in tracks and their human and animal traffic from many different directions to cross the river at that particular point, like so many threads grasped tightly in the hand. Patterns of fields, houses, farms, and so on, come to be configured in relation to the roads that radiate into and out from the crossing-place. In many cases, then, the landscape has been 'gathered', in Heidegger's sense of the word, around that particular crossing-point, long before the bridge was built.

But we have seen that fords can shift and bridges can be destroyed by flood or left high and dry by shifting river courses, bringing about a wrenching and warping of the surrounding landscape as it undergoes – to paraphrase Heidegger – a *re-gathering* around the new crossing-place.

Many archaeological landscapes have been ruptured in this way. In the following example, Biggleswade is a small town of Saxon origin named after the 'wade' or ford across the River Ivel on which it was founded, and around which it developed. But the location of the ford has been forgotten, and the flows of people and animals that once used it forced to seek other routes. Centuries ago the ford was replaced by a new bridge about 1 kilometre to the north of the town, pulling roads towards it away from the old crossing. Paths of movement leading up to the former ford have long since been blocked, their material traces overlaid and hidden by later features.

Looking at the landscape on an Ordnance Survey map today, the River Ivel seems to divide rural landscape from urban townscape. There is no apparent connection between the town and the enigmatic monument on the other side of the river, discovered from aerial photos in the 1930s. Although classified as a ringwork of Norman date, a very shallow archaeology was encountered in the only excavation of the site, with traces of timber and wattle-and-daub structures dated by Saxo-Norman pottery. The curious segmentation of the supposed ringwork ditches – clearly visible on aerial photos but never adequately explained – casts doubt on the accepted interpretation (Edgeworth 2008: 22-3).

Now consider the same landscape following the discovery by Cambridge Archaeological Unit of a broad droveway, thought

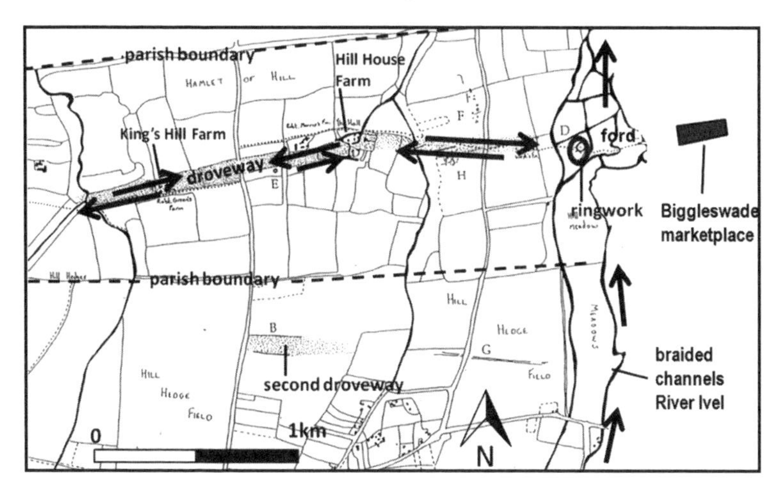

Fig. 5.1. A landscape reconnected: Anglo-Saxon droveway (base map by Mortimer and McFadyen 1999, Fig. 12, reproduced and added to by permission of the Cambridge Archaeological Unit).

to be of Middle to Late Anglo-Saxon date, crossing the valley from west to east (Mortimer and McFadyen 1999: 48-59). The droveway was exceptionally wide, being defined by two parallel ditches 70 metres apart. It ran through an unusual 'corridor' of parallel parish boundaries about 1 kilometre apart, and in this respect is remarkably reminiscent of the landscapes of the *tratturi* or *cañadas*. Not only does the course of the droveway point to the location of the former ford, where it crosses the river, it also provides a connecting path of flow between the town and the ringwork. It goes directly past the ringwork on one side of the river and lines up exactly with the broad funnel-shaped Biggleswade market-place on the other.

The breadth of the droveway indicates the size of herds driven along it and through the ford. It must have been a spectacular and noisy scene when a herd was taken through the braided channels of the Ivel – mud and water splashing everywhere, thin ice splintering, currents making new eddies and whorls as animal bodies entered the stream, steamy clouds of perspiration arising from backs of beasts swimming or wad-

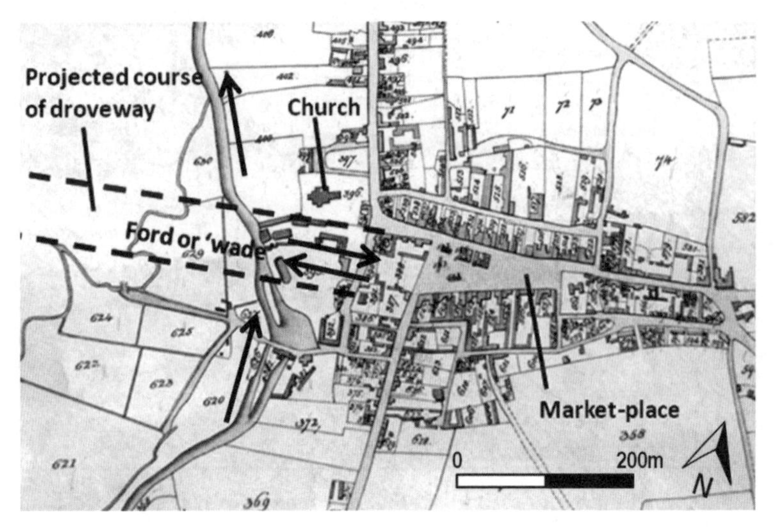

Fig. 5.2. A landscape reconnected: town and market-place (tithe map of Biggleswade 1838 used as base plan, by courtesy of Bedfordshire and Luton Archive and Record Service).

ing across, frosty breath from their nostrils, the calls of herders mixed in with the sounds of excited animals. It was a place of transformation between different states of matter, at the point where land and river flows intermeshed.

Almost every element of the landscape (market-place, town, ford, river, ringwork, parish boundaries, etc.) can now be radically re-interpreted in the light of interconnections with all the other elements. Previously thought to have been the result of late twelfth-century planning, the broad market-place is clearly directly on the line of what appears to be a much earlier linear feature (the droveway/ford). The course of the market-place *is* the course of the droveway. This invites a new theory of the origin and development not only of the market-place but of the town itself. It could be argued that the funnel-like shape of the market-place – the central space around which the layout of the town is based – is itself a structure of flow, at once shaping and shaped by the flowing energies that nourished urban development.

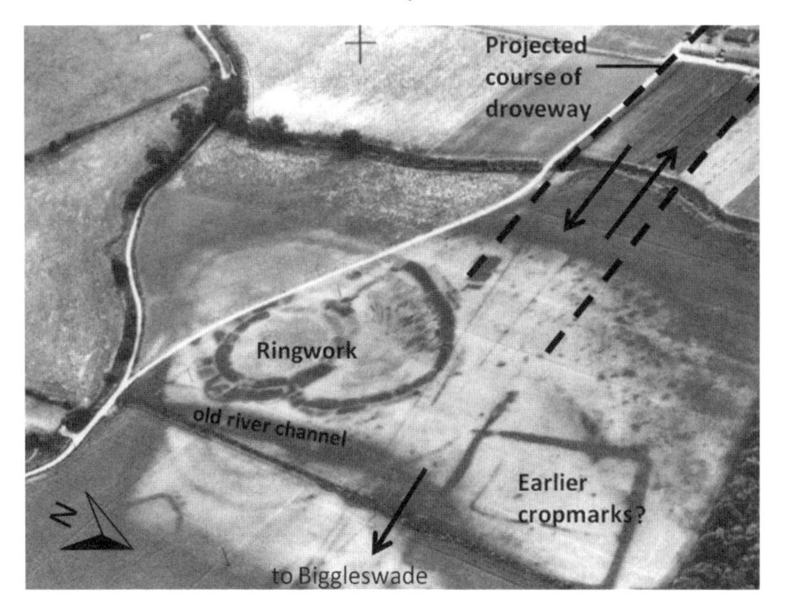

Fig. 5.3. A landscape reconnected: the 'ringwork' (1953 aerial photo, University of Cambridge Collection of Aerial Photography).

The so-called 'ringwork' located on the other side of the river (but formerly on a gravel island between braided channels) is also subject to reinterpretation in the light of the new discovery. It is no longer an isolated monument, separate from the town. It can now be seen to be directly on the route of the broad droveway at the very point where it channelled vast herds of animals across the river. The spiralling vortex-like form of the site brings to mind many other structures associated with flow of one kind or another. The way the spiral arm of the ringwork curves outwards to join the lines of droveway suggests channelling of flow from droveway to ringwork and vice versa. The same curving spiral arm also loops through a former river channel, and it is tempting to suggest that animals were being led through the stream as part of their journey into and out of the site (though excavation would be required to establish whether the different elements were contemporary). Standard interpretation of the monument as a Norman ringwork disregards completely the unusual segmentation pattern visible in the 'ditches' and other anomalous evidence. An alternative view of the site in terms of its affordances for flow, however, might suggest that such patterns represent structures associated with enclosure, corralling and sorting of animals. The site may not be an exact replica of an Icelandic *rétt*, but could be something very like it in terms of general function – a place for seasonal gathering or assembly of people and animals at the point where land and water flows meet. Whatever the site is, it seems probable that it was connected with control of animal movement along the droveway, through the broad corridor of parish boundaries, into and out of the incipient town.

The case of the ringwork serves as a useful example of the kind of 'gestalt switch' that can be initiated by the perception of landscapes in terms of flowing movement. Look at it first of all as an isolated monument, unconnected to the town on the other side of the river, set within a landscape relatively devoid of flow. Thus conceived, it falls readily into our established system of site classification; we place it in the same category as other castle-like monuments on the basis of superficial surface resemblances, notwithstanding anomalous evidence such as segmentation of ditches. Now look at it again in terms of flow – not just flow of the river but also the flow of large herds of animals channelled along the droveway and through the muddy waters of the ford (a crossing or intermeshing of flows). When seen as part of those currents, the site is re-animated, re-configured, taking on alignments in space in relation to the principal axes of flow (the river and the droveway). Our attention is directed to those extraordinary aspects of the site that were formerly neglected. We begin to see, furthermore, how moving animal and human bodies might themselves acquire an alignment and direction of movement in relation to the physical structure of the site that directs their motion. Meanings that were formerly fixed become fluid once more. Radical reinterpretation becomes possible (for more detailed analysis of this landscape, see Edgeworth 2008).

Summary

This chapter has considered to what extent the artefacts, sites, buildings and layouts of land-based archaeology can be perceived and interpreted in terms of eddies, currents and flows. Many aspects of past human activity, from gift exchange to mortuary practices, movement of people to herding of animals, can be understood in terms of 'flows of substances' or 'flows of materials'. It was noted in particular that material traces of flows of people and animals, in the form of roads and tracks, are intermeshed with those of river flows to some extent. Places where different kinds of flow cross over or merge into each other, such as fords and bridges, are key places in landscapes of flow. Sites and monuments can be interpreted with regard to their affordances for flow. Considerations of flow can potentially transform interpretation of otherwise fragmented and disconnected landscapes.

6

Contemporary and future entanglements

Introduction

This book has been concerned with past entanglements with rivers and other kinds of material flow. Now it is time to acknowledge that the practice of archaeology today is inextricably caught up with rivers and those other flows too, and to briefly explore the role of archaeology in contemporary water politics. Flowing water is increasingly a contested resource in many parts of the world. With global population rising from one billion to nearly seven billion in the last two hundred years, and corresponding growth of mega-cities, tremendous pressure has been placed on rivers and associated distributary networks for drinking, irrigation and industrial use – especially for the provision of hydro-electric power.

Since the 1940s, the number and size of large dams has risen dramatically. There are presently nearly 50,000 dams over 15 metres high on the world's rivers – amounting to a massive intervention in water flow and sediment flux on an unprecedented scale, not to mention river ecology. There are over 300 dams higher than 150 metres. Nearly half a million square kilometres have been flooded by dams (see Nilsson et al. 2005 for a global overview of the impact of dams on rivers), and it is thought that up to 50 million people have had to be resettled to other areas as a result. The Three Gorges dam on the Yangtze River in China has submerged over a hundred towns and upwards of a thousand villages (not to mention archaeological sites).

Yet a significant proportion of the global population still does not have direct access to basic necessities like water and electricity. Power and resources are distributed unequally. Many rivers are heavily polluted by industrial residues, agricultural chemicals and sewage from further upstream. Water is a contested resource both within and between countries through which rivers flow. Thus the geographer Erik Swyngedouw writes cogently of the 'flows of power' which contribute to the situation in the modern city of Guayaquil, Ecuador. Despite the location of the city on the confluence of two large rivers, over half a million slum dwellers there have no direct access to drinkable water (Swyngedouw 2004).

Archaeology as politically engaged with flowing water

It has already been noted that dams have extensive effects both upstream and downstream of the point of intervention. This applies not only to patterns of flow, erosion and deposition of sediment. It also applies to social and economic well-being of people who live alongside rivers. For those living upstream on land due to be flooded, loss of territory and forced resettlement is often the stark reality. For those living downstream in the shadow of the dam, there may be drying out of wetlands, radical changes in river flow, and loss of resources and livelihoods. Migratory fish flows are disrupted and the whole ecological balance of the river gets completely thrown (Palmer 2004). Such negative impacts have to be balanced, of course, against positive effects on the lives of people in communities and cities that are part of the wider networks served by the dam. Hydroelectric power plants produce nearly a fifth of electricity worldwide.

What does this have to do with archaeology? Archaeologists are often contracted to survey, excavate and record the cultural heritage of the land to be flooded, sometimes even to help physically move monuments out of the flood zone. When flooding of a river valley is resisted by local people, and decisions of governments and multinational companies are contested, archaeologists are inevitably faced with ethical dilemmas. Van de Noort and O'Sullivan cite the case of the building of dams on the upper reaches of the Tigris and Euphrates Rivers in Turkey, impacting on the lives of Kurdish people who lived along river systems and associated wetlands. The roles played by archaeologists in that complex political context are discussed, and it is pointed out that archaeologists not only carried out much valuable work in saving cultural heritage but also aligned with local people and activists in a partially successful campaign against European funding for the key Ilisu dam (Van de Noort and O'Sullivan 2006: 120-3; Ronayne 2006). Drawing from that experience, Maggie Ronayne argues that archaeologists in such circumstances have a duty both to protect or salvage cultural heritage *and* to be accountable to local communities (Ronayne 2005, 2007).

An example of where archaeologists allegedly failed to align themselves sufficiently with interests of local people is provided in a powerful recent paper by Henriette Hafsaas-Tsakos. The building of the Merowe Dam on the Fourth Cataract of the River Nile in Sudan led to forced resettlement of up to 78,000 people, including almost the entire Manasir tribe. Archaeologists were partially implicated in that compulsory migration, even if not responsible for it. Archaeological missions from a range of countries were contracted to work on salvaging the cultural heritage from affected regions. But relations between archaeologists and the Manasir, initially good, started to deteriorate. According to Hafsaas-Tsakos, the Manasir started to perceive that the archaeological work legitimised both the government's work in building the dam and the forced resettlement, and that there was a flow of artefacts leaving Sudan for other countries. As a result, some of the archaeological missions were expelled from Manasir tribal land (Hafsaas-Tsakos 2011).

This is not a book about ethics and it is not intended to pass judgement on difficult decisions made in ethically challenging circumstances, merely to point out that these examples show the very act of doing archaeology to be an inherently political act, inextricably embedded in wider systems and networks, and that there is an ethical dimension to what we do. It is especially the case on rivers, where divisions and disputes tend to be configured around the material axis of upstream and downstream and the direction of water flow, as well as the basic human requirements for water and energy – adding extra layers of complexity to the political landscape.

Even in temperate countries like Britain, where there is less pressure on water supplies, archaeologists are entangled with rivers in a socio-economic sense. Much of the funding for archaeological excavation comes from developers and extractors. Extraction of minerals like gravel and clay from floodplains, building of bridges and embankments, and construction of housing estates in low-lying areas of river valleys, have radical effects for better or worse on river flow and therefore on the lives of people living in floodplains, not only in the present but in the future too. Although we may like to regard ourselves as objective investigators interested only in the past, taking a neutral or distanced stance on contemporary issues, it has to be acknowledged that here too that the practice of archaeology is part of wider networks and processes, and as such is inextricably tangled up with river and floodplain change.

Enmeshed in broader structures of power, the voice of archaeology can however be a very important one, not least in furnishing examples of forms of past river modifications that have great relevance for creating sustainable river technologies in the present and into the future. There are many useful lessons to be learnt from past successes and mistakes in modification and use of flowing water. The archaeological record is full of material traces of the most extraordinary and ingenious systems and structures, from networks of underground qanats in the Middle East to linked channels and reservoirs that sustained the city and temples of Angkor in Cambodia, which have some relevance for the working out of viable ecological strategies today. Inspiring instances of river installations include the Brewarrina fish-traps in Australia, which combine spiritual values with practical expertise and knowledge of the environment: such material structures bring communities together rather than divide them, and are not used to exploit resources to maximum capacity or exhaust migratory flows of fish.

That archaeology has a role to play in changing ideas about rivers in the present as well as in the past is shown by the way in which major flooding disasters are invariably characterised as 'natural' disasters – as though rivers have no cultural or historical dimension that might need to be included in the explanation. The idea of the river as a 'natural' entity has a strong hold in the popular as well as the academic imagination. Thus unusual monsoon patterns and exceptionally heavy rainfall were cited by the media as principal reasons for the devastating floods in Pakistan in 2010. While this is partially true, a more in-depth picture is gained by looking for causes in the entanglement of natural and cultural forces. For example, rapid deforestation brought about by logging within the River Indus catchment area – increasing soil erosion and groundwater run-off – was an important factor. Stacks of logs stored in countless narrow ravines for smuggling by 'timber mafia' were dislodged in the floods, to be carried by the torrent and sent hurtling into dams and levees and bridges further downstream (Shamsie 2010). Huge logjams weakened river control structures and created bottlenecks of flow that greatly worsened the effects of the flooding. However, little discussion of the relationship between illegal logging and flooding appeared in the news.

It is important to realise that the River Indus and its tributaries constitute one of the most shaped and modified river systems in the world. With part of its watershed in India and part in Pakistan, it is also one of the most politicised. Over 150 million people are dependent on the river for drinking water, irrigation and hydroelectric power today. The river is diverted and extended far beyond its tributary structure through a vast distributary network of dams, barrages, reservoirs, levees, canals and irrigation channels – the earliest elements of which can be traced to the hydraulic engineering of the Harappan civilisation and other societies of the Bronze Age. As an assemblage of cultural and natural forces the Indus has a complex developmental history, with any 'natural' flow regime it may have once had long since replaced by flow regimes that are part natural, part cultural. In this context, and given numerous comparative examples available from other heavily modified river systems throughout the world, it would perhaps be foolish to attempt to explain or deal with any aspect of flow – including present-day flooding events – without considering the human entanglement with the river and the material legacy left by the past.

In this chapter our perspective has shifted from looking back at rivers and other material flows in the past to looking at the present and forward into the future. In that spirit, it seems appropriate to finish the book not so much with the usual conclusion and summing up, and the backward glance it implies, but rather with a more forward-looking manifesto for an archaeology of flow. The manifesto builds upon the more detailed examples provided in the book, but also stands in its own right and can be taken as a distillation of the argument of the book as a whole.

Manifesto for archaeology of flow

Matter can be in any one of three main states: solid, liquid or gas. In the archaeological study of landscapes, solid matter takes priority. Pick up almost any book on British landscape archaeology and you will find solid materials highlighted, with flowing liquid and gaseous materials cast into shadow. Rivers and streams are the dark matter of landscape archaeology (but no less vibrant for all that). Running through the heart of landscapes, shape-shifting and state-changing as they go, they are rarely subjected to the kind of cultural analysis applied to solid materials. Flowing water tends to be regarded as part of a natural background *against* which past cultural activity shows up, *next to* which sites are located, *onto* which cultural meaning is applied or *into* which cultural items are placed, rather than having any cultural dimension in its own right. Yet human activity, in the form of modification of rivers, is inextricably bound up with the so-called 'natural' water cycle. As dynamic entanglements of natural and cultural forces, rivers have the potential to re-shape landscape and our understanding of it. This manifesto presents six interlinked reasons for bringing the dark matter of landscapes into the domain of archaeological study.

1. Rivers are cultural artefacts

Rivers, especially in densely populated countries like Britain, are some of the most culturally modified of all landscape features. But in using the term artefact, I do not just mean that rivers and their flow have been artificially shaped. I also mean that, in being manipulated and controlled to some extent, their flow is used to shape other things. Through watermills, flow was deployed in the past to shape numerous materials and turn these into artefacts too. More recently, electricity generated from hydro-electric power plants on rivers has been turned to countless uses in shaping every aspect of the modern industrialised world. River flow has even been utilised in wartime as a weapon. Modified and manipulated rivers have also gone on to change the shape of deltas, floodplains, and other large-scale landforms.

2. Rivers are partially wild

No matter how shaped, controlled and managed they may be, rivers also have a wild aspect that is not entirely predictable, can act in unforeseen and surprising ways, and have the capacity at least temporarily to escape from culturally applied forms. That wildness means that any attempt to control flow will not simply be the application of a cultural force onto an inert and passive substance, for flowing water is an especially vibrant kind of matter that can act or respond in sometimes unforeseen and surprising ways, requiring counter-responses. It makes any human involvement with rivers more like a wrestle, an intertwining, a confluence, an enmeshment, an assemblage or an entanglement. Whatever metaphor we use, it is this dynamic merger of natural and cultural materials and agencies, ravelling and unravelling through time, that makes the archaeological study of rivers so interesting.

3. Human activity and river activity are intertwined

It used to be assumed that river activity and floodplain formation were mainly natural processes, therefore not subject to archaeological (cultural) analysis. But it turns out that many of the standard hydrological models of river erosion and sedimentation are based on studies of streams that – far from being natural as thought – had actually been subject to extensive human modification in the past. Evidence of extensive human intervention in river and floodplain morphology is clear for the modern world, not so obvious for earlier periods. Yet it can be found, for example, in medieval Europe and along the wadis of the ancient Near East, or the monumental levees and raised floodplains of the Yellow River in China. For their part rivers have woven their way into the very fabric of human existence – flowing through the centre of towns, under bridges, beside parks and gardens, into sluices and culverts and cooling towers. Rivers also run through dreams, songs, designs, projects, poems, memories and myth. They are part of the human story.

4. Understanding rivers entails understanding past human activities (and vice versa)

Now is the time to do away with those old physical geography lessons and ubiquitous diagrams that present the hydrological cycle (evaporation → condensation → precipitation → flow → evaporation → and so on) as entirely natural processes, somehow separate from human activity. In intervening in patterns of river flow – either directly (through damming, diversion, dredging, embanking, draining, irrigation, etc.) or indirectly (through deforestation, agricultural practices, etc.) – humans have been a part of the water cycle for thousands of years, effecting sediment flows and landscape formations. Rivers and streams have long been cyborgs (Haraway 1985) or hybrids

(Latour 1993) – dynamic assemblages of materials, flows and forces, both human and non-human. Human interventions in rivers today are of a much greater order of magnitude, it is true, but these are still on historical trajectories of human-river entanglement originating in the more or less distant past. It might well be asked, how can rivers be understood, and how can effective strategies be put in place for dealing with rivers, if those historical trajectories are not taken into account?

5. Rivers are dangerous, therefore good to think with

As when a river in flood breaks through or over its artificial banks, and carves itself a new channel, flow always threatens to break down the cultural order of things. It is precisely this dangerous and wild aspect of rivers that makes them good to think with. Flow has its own logic, which works in eddies, currents, streamlines, vortices and turbulences, flowing round and over the logic of non-flowing solid materials. It encourages us to break down polarities of thought, such as rigid oppositions between nature and culture, and not to respect too much the boundaries between different disciplines. Adopting a multi-disciplinary approach, making use of insights from both natural science and cultural studies, shifting between scales of analysis, looking always for different ways of looking at things, would be entirely in keeping with archaeology of flow. Flow itself challenges us to adopt more fluid and dynamic forms of investigation. To think in terms of flow leads to a greater emphasis on continuities – less on discontinuities. Simply by bringing flow into the scope of study has the potential to change our way of thinking about things radically.

6. Flowing water provides models for understanding other kinds of landscape flows

Water is not the only kind of material that flows through archaeological landscapes. People, goods, money, vehicles, animal herds, and many other entities exhibit flowing patterns of behaviour, leaving traces in the archaeological record. Nor are rivers and streams the only kind of material feature to channel flow. Paths, hollow-ways, processional routes, staircases, station concourses, signposts, high street banks, fibre-optic cables, turnstiles at football grounds, layouts of streets within a town or city, and so on, all channel material flows of one kind or another, one of these flows being the movement of archaeologists themselves. Even the painted animals in the caves of Lascaux have a flow to them, when considered in the light of the perspective of an embodied perceiver moving through the caves, as opposed to studying them from a fixed standpoint.

What happens if we apply models of flow to archaeological evidence that has previously been understood only as solid material?

Bibliography

Abizaid, C. (2005) 'An anthropogenic meander cutoff along the Ucalayi River, Peruvian Amazonia', *Geographical Review* 95, 1: 122-35.

Aldred, O. (2009a) 'Going along, remembering the way: an archaeology of movement in Iceland' (Stanford University Metamedia website), accessed 03.07.2010, http://humanitieslab.stanford.edu/107/3376

Aldred, O. (2009b) 'Réttir in the landscape: archaeological investigations of sheepfolds in Skútustadarhreppur and other neighbouring districts', accessed 03.07.2010, http://www.nabohome.org/publications/ipy/Rettir_in_the_landscape_ipy_report_2009_sm.pdf

Aldred, O. and Madsen, C.K. (2007) 'Rettir in the landscape: a study on the interactions between humans and animals through sheepfold monuments', accessed 03.07.2010, http://www.nabohome.org/publications/ipy/Rettir_in_the_landscape_ipy_report.pdf

Aston, M. (ed.) (1988) *Medieval fish, fisheries and fishponds in England*, vols 1 and 2 (Oxford: BAR Reports).

Ball, P. (2009a) *Flow* (Oxford: Oxford University Press).

Ball, P. (2009b) *Branches* (Oxford: Oxford University Press).

Barker, G. (2002) 'A tale of two deserts: contrasting desertification histories on Rome's desert frontiers', *World Archaeology* 33,3: 488-507.

Barton, N. (1992) *The lost rivers of London*, revised edn (London: Historical Publications Ltd).

Batey, C.E. and Morris, C.D. (1992) 'Earl's Bu, Orphir, Orkney: excavation of a Norse horizontal mill', in C.D. Morris and D.J. Rackham (eds) *Norse and later settlement and subsistence in the North Atlantic* (Glasgow: University of Glasgow): 33-41.

Batey, C.E. (1993) 'A Norse horizontal mill in Orkney', *Review of Scottish Culture* 8: 20-8.

Beekman, C.S., Weigand, P. and Pint, J. (1999) 'Old World irrigation technology in a New World context: qanats in Spanish colonial western Mexico', *Antiquity* 73, 280: 440-6.

Bender, B. (2001) 'Landscapes on-the-move', *Journal of Social Archaeology* 1: 75-89.

Bennett, J. (2010) *Vibrant matter: a political ecology of things* (Durham, Duke University Press).

Blair, J. (ed.) (2007) *Waterways and canal-building in medieval England* (Oxford: Oxford University Press).

Bond, J. (2001) 'Monastic water management in Great Britain: a review', in G. Keevill, M. Aston and T. Hall (eds) *Monastic Archaeology* (Oxford: Oxbow): 88-136.

Bonnamour, L. (2000) *Archéologie de la Saône* (Paris: Editions Errance).

Bradley, R. (1990) *The passage of arms* (Cambridge: Cambridge University Press).

Bradley, R. (2000) *An archaeology of natural places* (London: Routledge).

Brooks, N.P. (1966) 'Excavations at Wallingford Castle, 1965: an interim report', *Berkshire Archaeological Journal* 62: 17-21.

Brophy, K. (1999) 'Seeing the cursus as a symbolic river', *British Archaeology* 44: 6-7.

Brown, A. (1997) *Alluvial geoarchaeology: floodplain archaeology and environmental change* (Cambridge: Cambridge University Press).

Brown, A. (2009) 'The geomorphology and environment of the Hemington reach', in Ripper and Cooper (eds): 142-73.

Brown, G. (2005) 'Irrigation of water meadows in England', in Klapste (ed.): 84-92.

Brück, J. (2005) 'Experiencing the past? The development of a phenomenological archaeology in British prehistory', *Archaeological Dialogues* 12: 45-72.

Buteux, S. and Chapman, H. (2009) *Where rivers meet: the archaeology of Catholme and the Trent-Tame confluence* (York: Council for British Archaeology).

Candy, J. (2005) 'Landscape and perception: the medieval pilgrimage to Santiago de Compostela from an archaeological perspective', *eSharp 4*, accessed 03.03.2011, http://www.gla.ac.uk/media/media_41150_en.pdf

Candy, J. (2009) *The archaeology of pilgrimage on the Camino de Santiago de Compostela: a landscape perspective*, BAR International Series 1948 (Oxford: Archaeopress).

Certeau, M. de (1984) *The practice of everyday life,* tr. S. Rendall (Berkeley: University of California Press).

Christie, N. (2008) 'Of sheep and men; castles and transhumance in the upper Sangro Valley and in the Cicolano, Italy', in G. Lock and A. Faustoferri (eds) *Archaeology and landscape in Central Italy: papers in memory of John. A. Lloyd* (Oxford: University of Oxford School of Archaeology): 105-20.

Clark, C. (ed.) (1983) *Flood* (Amsterdam: Time-Life Books).

Clay, P. and Salisbury, C.R. (1990) 'A Norman mill dam and other sites at Hemington Fields, Castle Donington, Leicestershire', *Archaeological Journal* 147: 276-307.

Clifford, J. (1992) 'Travelling cultures', in L. Grossberg, C. Nelson and P. Treichler (eds) *Cultural studies* (New York: Routledge).

Coles, B.J. (2006) *Beavers in Britain's past* (Oxford: Oxbow).

Coles, B.J. and Coles, J.M. (1989) *People of the wetlands: bogs, bodies and lake-dwellers* (London: Thames and Hudson).

Coles, B.J. and Lawson (eds) (1987) *European wetlands in prehistory* (Oxford: Oxford University Press).

Cooper, A. (2006) *Bridges and power in medieval England, 700-1400* (Woodbridge, Boydell Press).

Cooper, L.P. (2003) 'Hemington Quarry, Castle Donington, Leicestershire, UK: a decade beneath the alluvium in the confluence zone', in Howard, Macklin and Passmore (eds): 27-41.

Courtney, P. (2009) 'Crossing the Trent: the Hemington bridges in local and regional context', in Ripper and Cooper (eds): 174-220.

Creighton, O. and Higham, R. (2005) *Medieval town walls: an archaeology and social history of defence* (Stroud: Tempus).

Dargin, P. (1976) *Aboriginal fisheries of the Darling-Barwon Rivers* (Brewarrina: Brewarrina Historical Society).

Dodgen, R.A. (2001) *Controlling the dragon: Confucian engineers and the Yellow River in late Imperial China* (Minoa: University of Hawaii Press).

Downward, S. and Skinner, K (2005) 'Working rivers: the geomorphological legacy of English freshwater mills', *Area* 37, 2: 138-47.

Edgeworth, M. (2008) 'Linking urban townscape with rural landscape: evidence of animal transhumance in the River Ivel Valley, Bedfordshire', *Medieval Settlement Research* 23: 22-7.

Edgeworth, M. (2009) 'Comparing burhs: a Wallingford-Bedford case-study', in K. Keats-Rohan. and D. Roffe (eds), in *The origins of the borough of Wallingford: archaeological and historical perspectives*, BAR British Series 494 (Oxford: Archaeopress): 77-85.

Edgeworth, M. and Christie, N. (2011) 'The archaeology of crossing-places', in Bayerische Gesellschaft für Unterwasserarchäologie e. V. (ed.) *Archäologie der Brücken – Vorgeschichte, Antike, Mittelalter, Neuzeit* (Regensburg: Verlag Friedrich Pustet).

Edgeworth, R.L. (1820) *Memoirs of Richard Lovell Edgeworth,* vol. 1 (London: R. Hunter).

English, P.W. (1968) 'The origin and spread of qanats in the Old World', *Proceedings of the American Philosophical Society* 112: 170-81.

Erickson, C. (2000) 'An artificial landscape-scale fishery in the Bolivian Amazon', *Nature* 408: 190-3.

Fahlbusch, H. (2009) 'Early dams', *Engineering History and Heritage* 162: 13-18.

Fisk, H.N. (1944) *Geological investigation of the alluvial valley of the lower Mississippi River*, Report for the US Army Corps of Engineers, Vicksburg, MS.

Fleming, A. (1999) 'Phenomenology and the megaliths of Wales: a dreaming too far?' *Oxford Journal of Archaeology* 18, 2: 119-25.

Fortune, M. (1988) 'Historical changes of a large river in an urban area: the Garonne River, Toulouse, France', *Regulated Rivers: Research and Management* 2, 2: 179-86.

Foster, B.R. (ed.) (2001) *The Epic of Gilgamesh*, tr. B.R. Foster (New York: Yale University Press).

Fowden, G. (1999) 'Desert kites: ethnography, archaeology and art', *Journal of Roman Studies*, Supplementary Series 31: 107-36.

Fowler, C. (2004). *The archaeology of personhood: an anthropological approach* (London: Routledge).

Gaimster, David (2009) *International handbook of historical archaeology* (New York: Springer).

Gimpel, J. (1992) *The medieval machine: the industrial revolution of the Middle Ages*, 2nd edn (London: Pimlico).

Goldsworthy, A. (2007) *Enclosure* (London: Thames and Hudson).

Gómez-Pantoja, J. (2004) 'Pecora consectari: transhumance in Roman Spain', in B. Santillo Frizell (ed.) *Pecus: man and animal in antiquity* (Rome: The Swedish Institute): 94-102.

Gräf, D. (2006). *Boat mills in Europe from early medieval to modern times,* tr. Michael Harverson and Leo van der Drift (Dresden: TIMS).

Graves-Brown, P. (ed.) (2005) *Matter, materiality and modern culture* (London: Routledge).

Graves-Brown, P. (2007). 'Concrete islands', in L. McAtackney, M. Palus and A. Piccini (eds) *Contemporary and historical archaeology in theory: papers from the 2003 and 2004 CHAT conferences*, BAR International Series 1677 (Oxford: Archaeopress): 75-82.

Grayson, A. (2004) 'Bradford's Brook, Wallingford', *Oxoniensia* 69: 29-44.

Guillerme, A. (1988) *The age of water: the urban environment in the north of France, AD 300-1800* (College Station: Texas A & M University Press).

Haevaert, V. and Baeteman, C. (2008) 'A middle to late Holocene avulsion history of the Euphrates River', *Quaternary Science Reviews* 27: 2401-10.

Hafsaas-Tsakos, H. (2011) 'Ethical dimensions of salvage archaeology and dam building: the clash between archaeologists and local people in Dar al-Manasir, Sudan', *Journal of Social Archaeology* 11, 1: 49-76.

Haraway, D. (1985) 'A manifesto for cyborgs: science, technology and socialist-feminism in the 1980s', *Socialist Review* 80: 65-108.

Harrison, D. (2004) *The bridges of medieval England: transport and society 400-1800* (Oxford: Clarendon Press).

Haughey. F. (1999) 'The archaeology of the Thames: prehistory within a dynamic landscape', *London Archaeologist* 9: 16-1.

Hawkes, J., Cross, J., Fasham, P. and Carruthers, W. (1997) *Excavations on Reading waterfront sites 1979-1988* (Salisbury: Trust for Wessex Archaeology).

Heidegger, M. (1971) 'Building, dwelling, thinking', in *Poetry, language, thought*, tr. Albert Hofstadter (New York: Harper and Row): 150-70.

Helms, S.W. and Betts, A. (1987) 'The desert kites of the Badiyat esh-Sham and Northern Arabia', *Paléorient* 13, 1: 41-67.

Heng, C.K. (1999) *Cities of aristocrats and bureaucrats: the development of medieval Chinese cityscapes* (Singapore: Singapore University Press).

Hicks, D. and Beaudry, M.C (eds) (2006) *The Cambridge companion to historical archaeology* (Cambridge: Cambridge University Press).

Hicks, D., McAtackney, L. and Fairclough, G. (eds) (2009) *Envisioning landscape situations and standpoints in archaeology and heritage* (Walnut Creek, CA: Left Coast Press).

Hindle, P. (2008) *Medieval roads and tracks*, 2nd edn (Colchester: Shire Publications).

Hodder, I. (1997) 'Always momentary, fluid and flexible: towards a reflexive excavation methodology', *Antiquity* 71, 273: 691-700.

Hodder, I. and Hutson, S. (2003) *Reading the past: current approaches to interpretation in archaeology*, 3rd edn (Cambridge: Cambridge University Press).

Hodge, T. (1992) *Roman aqueducts and water supply* (London: Duckworth).

Holtorf, C. (2008) Comments on M. Edgeworth, 'Rivers as artifacts', accessed 20.03.2011, http://traumwerk.stanford.edu/archaeolog/2008/04/rivers_as_artifacts_towards_an.html

Holzer, A., Avner, U., Porat, N. and Horwitz, L.K. (2010) 'Desert kites in the Negev desert and northeast Sinai: their function, chronology and ecology', *Journal of Arid Environments* 74: 806-17.

Howard, A.J., Brown, A.G., Carey, C.J., Challis, K., Cooper, L.P., Kincey, M. and Toms, P. (2008) 'Archaeological resource modelling in temperate river valleys: a case study from the Trent Valley, UK', *Antiquity* 82: 1040-54.

Howard, A.J. and Macklin, M.G. (1999) 'A generic geomorphological approach to archaeological interpretation and prospection in British river valleys: a guide for archaeologists investigating Holocene landscapes', *Antiquity* 73: 527-41.

Howard, A.J., Macklin, M.G. and Passmore, D.G. (eds) (2003) *Alluvial archaeology in Europe* (Lisse: Swets and Zeitlinger).

Ingold, T. (1993) 'The temporality of the landscape', *World Archaeology* 25,2: 24-174.

Ingold, T. (2000) *The perception of the environment: essays on livelihood, dwelling and skill* (London: Routledge).

Ingold, T. (2007) 'Materials against materiality', *Archaeological Dialogues* 14,1: 1-16.

Ingold, T. (2008) 'Bringing things to life: creative entanglements in a world of materials', accessed 03.03.2011, http://www.reallifemethods.ac.uk/events/vitalsigns/programme/documents/vital-signs-ingold-bringing-things-to-life.pdf

Jansen, R.B. (1980) *Dams and public safety* (Washington DC: US Department of Interior).

Jiongxin Xu (2003) 'Naturally and anthropogenically accelerated sedimentation in the Lower Yellow River, China, over the past 13,000 years', *Geografiska Annaler* Series A, 80,1: 67-78.

Johnson, L.C. (1995) 'China's Pompeii: twelfth century Dongjing', *Historian* 58: 49-68.

Johnson, M. (2006) *Ideas of landscape* (Oxford: Blackwell).

Kaufmann, C. (2007) 'A river's gifts', *National Geographic*, January issue: 150-7.

Klapste, J. (ed.) (2005) *Water management in the medieval rural economy*, Ruralia V, Pamatky Archeologicke – Supplementum 17 (Prague, Academy of Science of the Czech Republic).

Knappett, C. (2005) *Thinking through material culture: an interdisciplinary perspective* (Philadelphia: University of Pennsylvania Press).

Knighton, D. (1998) *Fluvial forms and processes: a new perspective* (London: Arnold).

Kowalik, P and Suligowski, Z. (2001) 'Comparison of water supply and sewage in Gdañsk (Poland) in three different periods', *Ambio* 30,4: 320-3.

Latour, B. (1993) *We have never been modern* (Cambridge MA, Harvard University Press).

Lawler, A. (2010) 'Uncovering a rural Chinese Pompeii', *Science* 328: 566-7.

Levi-Strauss, C. (1970) *The raw and the cooked*, tr. J. Weightman and D. Weightman. (London: Jonathan Cape).

Lewin, J. (2010) 'Medieval environmental impacts and feedbacks: the lowland floodplains of England and Wales', *Geoarchaeology* 25,3: 267-311.

Lillie, M. and Ellis, S. (ed.) (2006) *Wetland archaeology and environments. Regional issues, global perspectives* (Oxford: Oxbow Books).

Linton, J. (2008) 'Is the hydrological cycle sustainable? A historical-geographical critique of a modern concept', *Annals of the Association of American Geographers* 98,3: 630-49.

Lorge, P.W. (2005) *War, politics and society in early modern China, 900-1795* (London: Routledge).

Losey, R. (2010) 'Animism as a means of exploring archaeological fishing structures on Willapa Bay, Washington, USA', *Cambridge Archaeological Journal* 20,1: 17-32.

Luckhurst, D. (1964) *Monastic watermills: a study of mills within English monastic precincts* (London: Society for the Protection of Ancient Buildings).

Majewski, W. (2005) 'Learning to live with floods', *Academia* 1,5: 30-1.

Masters, R.D. (1999) *Fortune is a river: Leonardo Da Vinci and Niccolo Machiavelli's magnificent dream to change the course of Florentine history* (London: Simon and Schuster).

McPhee, J. (1989) *The control of nature* (New York: Farrar Straus Giroux).

McQuade, M. and O'Donnell, L. (2007) 'Late Mesolithic fish-traps from the Liffey estuary, Dublin, Ireland', *Antiquity* 81: 569-84.

Mei-e, R., and Xianmo, Z. (1994) 'Anthropogenic influences on changes in the sediment load of the Yellow River, China, during the Holocene', *The Holocene* 4: 314-20.

Mendieta, S., Marconis, R., Maisonabe, A. and Serres, M. (2010) *Le Bazacle: les noces de Toulouse et de la Garonne* (Toulouse: Editions Privat).

Milne, G. (2002) 'London's medieval waterfront', *British Archaeology* 68: 20-3.

Milne, G. and Hobley, B. (eds) (1981) *Waterfront archaeology in Britain and Northern Europe*, CBA Report 41 (York: CBA).
Mithen, S. (2010) 'The domestication of water: water management in the ancient world and its prehistoric origins in the Jordan Valley', *Phil. Trans. R. Soc.* 368, 1931: 5249-74.
Montgomery, D.R. (2008) 'Dreams of natural streams', *Science* 319, 5861: 291-2.
Mortimer, R. and McFadyen, L. (1999) *Investigation of the archaeological landscape at Broom, Bedfordshire, phase 4*, Cambridge Archaeological Unit, Report 320.
Muir, R. (2000) *The new reading the landscape: fieldwork in landscape history* (Exeter: University of Exeter Press).
Muir, R. and Muir, N. (1986) *Rivers of Britain* (London: Bloomsbury).
Nilsson, C., Reidy, C., Dynesius, M. and Revenga, C. (2005) 'Fragmentation and flow regulation of the world's largest river systems', *Science* 308: 405-8.
Normark, J. (2009) 'Outlining an archaeology of water', accessed 30.03.2011, http://haecceities.wordpress.com/2009/09/29/outlining-an-archaeology-of-water/
Oosthuizen, S. (1993) 'Saxon commons in South Cambridgeshire', *Proceedings of the Cambridge Antiquarian Society* 82: 93-100.
Oosthuizen, S. (2002) 'Ancient greens in "Midland" landscapes: Barrington, South Cambridgeshire', *Medieval Archaeology* 46: 110-15.
Palmer, T. (2004) *Lifelines: the case for river conservation*, 2nd edn (Lanham, MA: Rowman and Littlefield).
Patrick, C. and Rátkai, S. (2009) *The Bull Ring uncovered: excavations at Edgbaston Street, Moor Street, Park Street and The Row, Birmingham City Centre 1997-2001* (Oxford: Oxbow).
Phythian-Adams, C. (ed.) (1993) *Societies, cultures and kinship, 1580-1850: cultural provinces in English local history* (Leicester: Leicester University Press): 1-23.
Pryor, F. (1996) 'Sheep, stockyards and field systems: Bronze Age livestock populations in the Fenlands of eastern England', *Antiquity* 70, 268: 313-24.
Pryor, F. (2007) 'Beware the glutinous ghetto!', in M. Lillie and S. Ellis (eds) *Wetland archaeology and environments: regional issues, global perspectives* (Oxford: Oxbow).

Raffles, H. (2002) *In Amazonia: a natural history* (Princeton, Princeton University Press).

Raffles, H. (2003) 'Further reflections on Amazonian environmental history: transformations of rivers and streams', *Latin American Research Review* 38, 3: 165-87

Reader's Digest (1978) *Book of natural wonders: a guide to the world's most unforgettable places* (New York: Reader's Digest Association).

Reynolds, T.S. (1983) *Stronger than a hundred men: a history of the vertical water wheel* (Baltimore: John Hopkins University Press).

Rhodes, E. (2007) 'Identifying modification of river channels', in J. Blair (ed.): 133-52.

Ripper, S. and Cooper, L.P. (eds) (2009) *The Hemington bridges: the excavation of three medieval bridges at Hemington Quarry, near Castle Donington, Leicester* (Leicester: University of Leicester Archaeology Monographs).

Ronayne, M. (2005) *The cultural and environmental impact of large dams in southeast Turkey* (London, Kurdish Human Rights Project).

Ronayne, M. (2006). 'Archaeology against cultural destruction: the case of the Ilisu Dam in the Kurdish region of Turkey', *Public Archaeology* 5: 23-236.

Ronayne, M. (2007) 'The culture of caring and its destruction in the Middle East: women's work, water, war and archaeology', in Y. Hamilakis and P. Duke (eds), *Archaeology and capitalism: from ethics to politics* (California: Left Coast Press): 247-65.

Scarpino, P.V. (1997) 'Large floodplain rivers as human artifacts: a historical perspective on ecological integrity', Special report for the US Geological Survey.

Schofield, J. and Vince, A. (2003) *Medieval towns: the archaeology of British towns in their European setting* (London: Equinox).

Shamsie, K. (2010) 'Pakistan's floods are not just a natural disaster', *The Guardian*, 5 August, G2: 8.

Steward, J.H. (1955) *Irrigation civilizations: a comparative study* (Washington DC: Pan American Union).

Swyngedouw, E. (2004) *Social power and the urbanization of water: flows of power* (Oxford: Oxford University Press).

Tesch, F.W. (2003) *The eel* (Oxford: Blackwell).

Thomas, J. (1996) *Time, culture and identity: an interpretive archaeology* (London: Routledge).

Tilley, C. (1997) *A phenomenology of landscape: places, paths and monuments* (Oxford: Berg).

Tilley, C., Keane, W., Kuechler, S., Rowlands, M. and Spyer, P. (eds) (2006) *Handbook of material culture* (London: Sage).

Tóth, J.A. (2006) 'River archaeology – a new field of research', *Archeometriai Mühely* 1: 61-6.

Tsing, A. (2000) 'The global situation', *Cultural Anthropology* 15, 3: 327-60.

Twain, M. (1883) *Life on the Mississippi* (Boston: James R. Osgood).

UNESCO (1998) 'World Heritage List 872, Lyon (France)', accessed 10.10.2010, http://whc.unesco.org/archive/advisory_body_evaluation/872.pdf

Urry, J. (2007) *Mobilities* (Oxford: Polity Press).

Uzuka, K. and Tomita, K. (1993) 'Flood control planning – case study of the Tone River', *Journal of Hydroscience and Hydraulic Engineering* 4: 5-22.

Van de Noort, R. and O'Sullivan, A. (2006) *Rethinking wetland archaeology* (London: Duckworth).

Van Gennep, A. (1908) *Les Rites de Passage* (Emile Nourry: Paris).

Vita-Finzi, C. (1960) 'Post-Roman erosion and deposition in the wadis of Tripolitania', *Assoc. Internat. Hydrol. Scient.* 53: 61-4.

Walter, R.C. and Merritts, D.J. (2008) 'Natural streams and the legacy of water-powered mills', *Science* 319, 5861: 299-304.

Wessex Archaeology (2008) 'England's oldest bridge', accessed 13.04.2011, http://www.wessexarch.co.uk/projects/hampshire/testwood/ oldest_bridge.html

White, R. (1995) *The organic machine: the remaking of the Columbia River* (New York: Hill and Wang).

Wilkinson, T.J. (2003) *Archaeological landscapes of the Near East* (Tucson: University of Arizona Press).

Wilkinson, T.J. (2007) 'Ancient Near Eastern route systems: from the ground up', *ArchAtlas* 4, accessed 12.10.2010, http://www.archatlas.org/workshop/TWilkinson07.php

Wilkinson, T.J. and Rayne, L. (2010) 'Hydraulic landscapes and imperial power in the Near East', *Water History* 2, 115-44.

Wilson, A.I. (2006) 'The spread of foggara-based irrigation in the ancient Sahara', in D.J. Mattingly et al. (eds) *The Libyan desert: natural resources and cultural heritage* (London: Society for Libyan Studies): 205-16.

Witmore, C. (2007) 'Symmetrical archaeology: excerpts of a manifesto', *World Archaeology* 39, 4: 546-62.

Witmore, C. and Webmoor, T. (2008) 'Things are us! A commentary on human/things relations under the banner of a social archaeology', *Norwegian Archaeological Review* 41, 1: 53-70.

Wittfogel, K.A. (1957) *Oriental despotism: a comparative study of total power* (New Haven: Yale University Press).

Wohl, E. and Merritts, D. (2007) 'What is a natural river?', *Geography Compass* 1, 4: 871-900.

Wolman, M.G. and Leopold, L.B. (1957) 'River flood plains: some observations on their formation', *U.S. Geological Survey Professional Paper 282-C*, 87-107.

Woods, C. (2005) 'On the Euphrates', *Zeitschrift für Assyriologie* 95: 7-45.

Wulff, H.E. (1968) 'The qanats of Iran', *Scientific American*, April issue: 94-105.

Wylie, J. (2007) *Landscape* (London: Routledge).

Zalasiewicz, J. et al. (2008). 'Are we now living in the Anthropocene?', *GSA Today* 18:2, 4-8.

Index

LIBRARY, UNIVERSITY OF CHESTER